Inerter and Its Application in Vibration Control Systems

Michael Z. Q. Chen · Yinlong Hu

Inerter and Its Application in Vibration Control Systems

Science Press
Beijing

Springer

Michael Z. Q. Chen
School of Automation
Nanjing University of Science
and Technology
Nanjing, Jiangsu, China

Yinlong Hu
College of Energy and Electrical
Engineering
Hohai University
Nanjing, Jiangsu, China

ISBN 978-981-10-7088-4 ISBN 978-981-10-7089-1 (eBook)
https://doi.org/10.1007/978-981-10-7089-1

Jointly published with Science Press, Beijing, China
The print edition is not for sale in China Mainland. Customers from China Mainland please order the print book from: Science Press.

Library of Congress Control Number: 2019930569

This Springer imprint is published by the registered company Springer Nature Singapore Pte Ltd.
The registered company address is: 152 Beach Road, #21-01/04 Gateway East, Singapore 189721, Singapore

Preface

The topic of this book is a comprehensive introduction of inerter, a new two-terminal mechanical element proposed by Prof. Malcolm Smith from Cambridge University in 2002, and its recent advances in vibration control systems. One of the principal motivations for introducing inerter is due to a snag in the correspondence between passive mechanical and electrical networks, as the capacitor does not have a real corresponding mechanical element. Such a snag has been removed because of the invention of inerter, and as a result a spring–damper–inerter mechanical network can be directly transformed into an inductor–resistor–capacitor electrical network. In this way, the electrical network synthesis theory can be directly applied to the mechanical network synthesis. Based on this, much attention has been drawn to the inerter-based mechanical network synthesis, where the problem of how to realize a positive-real transfer function as specific mechanical networks while considering the simplicity, cost, and other realization requirements is of particular interest. Apart from the research interest on inerter-based mechanical network synthesis, inerter has been applied to a variety of mechanical systems, such as vehicle suspensions, train suspensions, buildings, motorcycle steering compensators, landing gears, wind turbines, etc. A common practice of applying inerters to a specific system is that inerter-based networks, usually more complex than the traditional networks without inerters, are given and then optimize the parameters of the inerter-based networks to achieve a better performance than the traditional ones. Such a procedure is effective to demonstrate the benefits of using inerters, but the drawback is that the basic function of an inerter for general vibration systems is concealed. Since all the application scenarios of inerter are a subset of general vibration systems, it demands a comprehensive investigation of inerters from a vibration system point of view, which is the main motivation of this book.

This book is intended to provide a comprehensive summary of recent results on inerter, and then introduce the recent advances by the author on the application of inerter in vibration systems. In Chap. 1, the concept of inerter, its physical realizations, and state-of-art applications are introduced. In Chap. 2, the influence of inerter on the natural frequencies of vibration systems is discussed. Then, in

Chaps. 3 and 4, the inerter-based vibration isolators and dynamic vibration absorbers are introduced, and the parameters' optimization methods are proposed. In Chap. 5, the inerter concept is extended to semi-active control by introducing the semi-active inerter concept, where physical embodiments of semi-active inerter are also discussed.

The book is intended as a text for graduate students and researchers. A certain basic level of knowledge of mechanical vibration, control theory, and optimization theory is a necessary prerequisite in order to follow the material presented here.

The authors are indebted to all those who have contributed to material presented in this book and whose identities can be deduced from our joint publications that appear in the bibliography.

Nanjing, China
June 2018

Michael Z. Q. Chen
Yinlong Hu

Contents

Chapter 1
Introduction

1.1 Inerter

Inerter is a new mechanical element proposed by Professor Malcolm C. Smith from Cambridge University, which is defined as *a mechanical two-terminal, one-port device with the property that the equal and opposite force applied at the terminals is proportional to the relative acceleration between the terminals* (Smith 2002a). The symbol of inerter is shown in Fig. 1.1, where in the notation of Fig. 1.1,

$$F = b(\dot{v}_2 - \dot{v}_1). \tag{1.1}$$

The constant b in (1.1) is called the *inertance* with the units of kilograms. The energy stored by the inerter can be quantified as $\frac{1}{2}b(v_2 - v_1)^2$.

The main motivation of proposing inerter is due to the incompleteness of the force-current analogy between mechanical and electrical systems. It is well known that the mechanical and electrical systems have very similar dynamics. For the force-current analogy between these systems, the force and velocity in mechanical systems can be analogized to the current and voltage in the electrical systems, respectively. In this way, the fixed reference point in an inertial frame, the kinetic energy, the potential energy in mechanical systems can be analogized to the electrical ground, the electrical energy, magnetic energy in electrical systems, respectively. The detailed force-current analogy is shown in Table 1.1. From the device point of view, the spring and damper in mechanical systems can be analogized to the inductor and resistor in electrical systems, respectively. However, it lacks a mechanical device to analogize the capacitor in electrical systems. Historically, the mass is seen as the mechanical element corresponding to the capacitor in electrical systems. However, from Newton's Second Law, the acceleration of the mass is relative to a fixed point in the inertial frame. This means the one terminal of the mass is the ground and the other terminal is the center of the mass. In other words, the mass is not a genuine two-terminal device. The electrical element that corresponding to the mass is actually a grounded capacitor. It is straightforward to sec that there exists performance

M. Z. Q. Chen and Y. Hu, *Inerter and Its Application in Vibration Control Systems*, https://doi.org/10.1007/978-981-10-7089-1_1

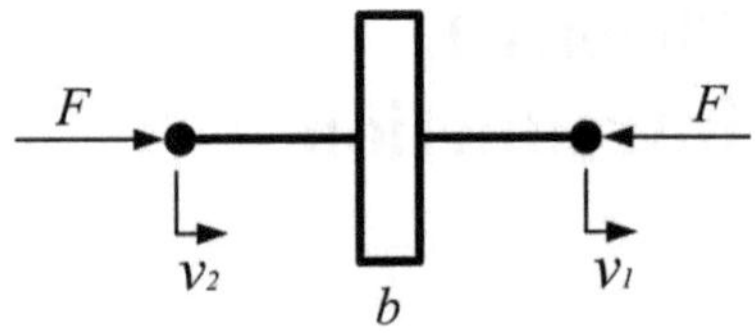

Fig. 1.1 The symbol of inerter

Table 1.1 The force-current analogy between mechanical and electrical systems

Mechanical systems	⇔	Electrical systems
Force	⇔	Current
Velocity	⇔	Voltage
Mechanical ground	⇔	Electrical ground
Kinetic energy	⇔	Electrical energy
Potential energy	⇔	Magnetic energy
Spring	⇔	Inductor
Damper	⇔	Resistor
Mass	⇔	Grounded capacitor

restrictions for the spring-damper-mass networks, as it is equivalent to the electrical networks composed of inductors, resistors, and capacitors but all capacitors should be grounded.

Due to the restriction between force-current analogy, the inerter is proposed. From the definition of inerter, one sees that inerter is a genuine two-terminal device which has similar dynamics with the capacitor. By proposing inerter, the force-current analogy is completed, and the spring-damper-inerter mechanical networks can be directly represented as inductor-resistor-capacitor electrical networks. A new correspondence after proposing inerter is shown in Fig. 1.2.

Note that although inerter is motivated by the force-current analogy, its properties are not dependent on this analogy, in the sense that, inerter has some unique functions for mechanical systems such as large equivalent mass (virtual mass), mechanical dual of the springs as an energy-storing element, etc. Such properties of inerter will be revealed in the following sections.

1.2 Network Synthesis

The introduction of inerter completes the analogy between the spring-damper-inerter mechanical networks and the inductor-resistor-capacitor electrical networks. Therefore, the systematic methods in passive electrical network synthesis can be directly applied to design inerter-based mechanical networks. In this part, some definitions and notations in electrical network synthesis are transformed into the mechanical domain by using the force-current analogy.

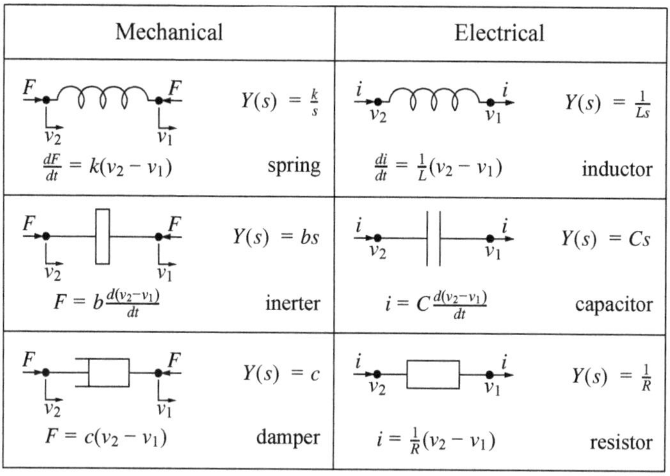

Fig. 1.2 The new correspondence between mechanical and electrical networks

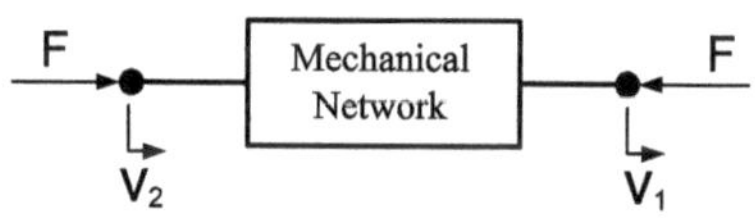

Fig. 1.3 A two-terminal mechanical network in a free-body form with force-velocity pair (F,v), where $v = v_2 - v_1$

Consider the two-terminal mechanical network shown in Fig. 1.3, the network is defined to be *passive* if for all admissible F and v, which are square integrable on $(\infty, T]$ (Anderson and Vongpanitlerd 1973)

$$\int_{-\infty}^{T} F(t)v(t)dt \geq 0. \tag{1.2}$$

The left-hand side of (1.2) indicates the total energy delivered to the network up to time T. Therefore, for a passive network, no energy is delivered to the environment.

Throughout this book, we define the *mechanical impedance* $Z(s)$ as the ratio between velocity and force, that is

$$Z(s) = \frac{\hat{v}(s)}{\hat{F}(s)},$$

where $\hat{\cdot}$ denotes the Laplace transform. The *mechanical admittance* $Y(s)$ is defined as the reciprocal of mechanical impedance $Z(s)$, that is

$$Y(s) = Z(s)^{-1}.$$

Here, the definitions of mechanical impedance and admittance are the same as the electrical circuits based on the force-current analogy. The force and current are seen as the through variables, and the velocity and voltage are seen as the cross variables. According to the traditional notation of electrical circuits, the impedance is defined as the ratio between cross variables and through variables.

The passivity of a network is related to the positive realness of the impedance and admittance of the network. For a two-terminal mechanical network, the network is *passive* if and only if its mechanical impedance or admittance is *positive real* (Smith 2002a). For a real-rational function $G(s)$ to be positive real, the sufficient and necessary conditions are Anderson and Vongpanitlerd (1973), Chen and Smith (2009)

1. $Z(s)$ is analytic and $Z(s) + Z(s)^* \geq 0$ in $Re(s) > 0$;
2. $Z(s)$ is analytic in $Re(s) > 0$, $Z(j\omega) + Z(j\omega)^* \geq 0$ for all ω, at which $Z(j\omega)$ is finite, and any poles of $Z(s)$ on the imaginary axis or at infinity are simple and have a positive residue.

The traditional electrical network synthesis indicates that any real-rational function $Z(s)$ which is positive real, there exists an inductor-resistor-capacitor electrical network whose impedance (or admittance) is $Z(s)$ (Bott and Duffin 1949). Based on the force-current analogy, it can be deduced that for any real-rational function $Z(s)$ which is positive real, there exists a spring-damper-inerter mechanical networks with an impedance (or admittance) as $Z(s)$ (Smith 2002a). Since (Chen and Smith 2009), there have been a series of new results for mechanical network synthesis (Chen et al. 2013, 2015; Wang et al. 2018).

1.3 The Physical Embodiments of Inerter

The definition of inerter introduces a new mechanical concept. The equally important issue for such a new concept is how to construct a real mechanical structure or device possessing the same or at less similar properties with the definition of inerter. The prodedure of constructing a physical embodiment of inerter is called realization.

Up to now, there are mainly three types of realizations of inerter, that is the rack-pinon inerter (Smith 2002a, b; Chen et al. 2009), the ball-screw inerter (Wang et al. 2009; Wang and Su 2008; Chen et al. 2009), and the hydraulic inerter (Wang et al. 2011; Gartner and Smith 2011; Tuluie 2010). Moreover, based on whether a flywheel is employed in the realization, it can classified into two categorizes, that is flywheel-based inerters (Smith 2002a; Wang and Su 2008; Chen et al. 2009; Wang et al. 2011), and non-flywheel inerters (Gartner and Smith 2011; Tuluie 2010).

Figure 1.4 shows the schematic and physical embodiments of a rack-pinion inerter (Wang and Su 2008; Chen et al. 2009). In this type of inerter, a plunger silding in a cylinder which derives a flywheel through a rack, pinion, and gears. If neglecting the masses of the plunger, the rack-pinion and gears, the dynamics of this structure can be approximated as

$$F = (m\alpha_1^2\alpha_2^2)\dot{v},$$

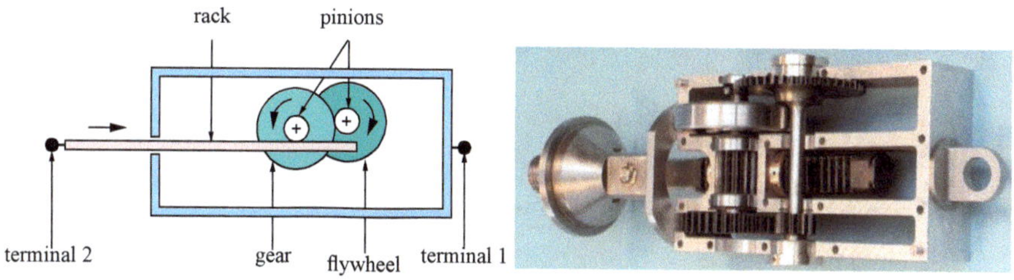

Fig. 1.4 The schematic of the rack-pinion inerter (Smith 2002a) (left) and the physical embodiment manufactured in the Cambridge University Engineering Department (CUED) workshops (Chen et al. 2009) (right)

where m is the mass of the flywheel, $\alpha_1 = \gamma/r_3$, $\alpha_2 = r_2/r_1$, with r_1, r_2, r_3, and γ denoting the radius of the rack pinion, the gear, the flywheel pinion, and the gyration of the flywheel. According to the definition, the inertance of the rack-pinion inerter is

$$b = m\alpha_1^2\alpha_2^2.$$

Clearly, if α_1 and α_2 are larger than 1, the inertance is lager than the flywheel mass m. This means that the rack pinion has the function of magnifying the mass of the flywheel. For example. if $\alpha_1 = \alpha_2 = 3$, where these quantities can easily be realized in practice, the inertance would be 81 times larger than the flywheel mass m. The right figure of Fig. 1.4 shows a physical embodiment of a rack-pinion inerter manufactured in the Cambridge University Engineering Department. The total mass of the structure is approximately 3.5 kg, while the realized inertance is about 725 kg (Smith 2003).

The rack-pinion inerter can bear a large load, and as a result, a large inertance can be realized. However, the inherent friction and the backlash between the gears significantly increase the nonlinearities of the inerter. A possible solution is to replace the rack pinion driving motion by a ball-screw type, then the second generation of the inerter, namely the ball-screw inerter, is proposed (Wang et al. 2009; Wang and Su 2008; Chen et al. 2009). Compared with the rack-pinion inerter, the friction is greatly reduced, and the backlash can be eliminated by pre-loading. Figure 1.5 shows a prototype manufactured in the Cambridge University Engineering Department, where the real mass is approximately 1 kg, while the realized inertance is about 180 kg (Smith 2011).

Similar to the rack-pinion inerter, the inertance of the ball-screw inerter can represented as the product of a transmission ratio and the flywheel's moment of inertia, that is

$$b = J\beta^2, \tag{1.3}$$

where J denotes the moment of inertia of the flywheel, β is the transmission ratio of the ball screw with $\beta = 2\pi/p$, p (in units of m/rev) is the pitch of the screw (Wang and Su 2008).

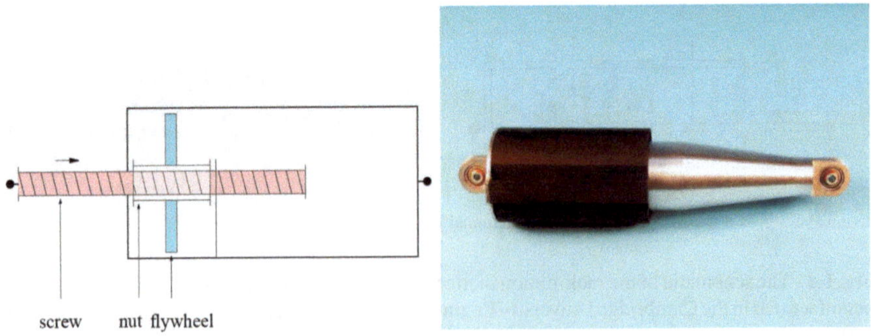

Fig. 1.5 The schematic of the ball-screw inerter (left) and the physical embodiment manufactured in the Cambridge University Engineering Department (CUED) workshops (right) (Chen et al. 2009)

Another flywheel-based inerter is the hydraulic inerter proposed in Wang et al. (2011), as shown in Fig. 1.6, where the hydraulic transmission motion is employed. The theoretical inertance can be calculated as

$$b = J\left(\frac{A}{D}\right)^2, \tag{1.4}$$

where J is the moment of inertia of the flywheel, A is the area of the piston, D is constant in the units of m^3 (Wang et al. 2011).

In summary, a common feature of the flywheel-based inerters is that a flywheel is driven by a transmission structure (the rack pinion, the ball screw, and the hydraulic type), and the moment of inertia of the flywheel is magnified by the transmission structure to realize the effect of the inerter. Therefore, there are two requirements for the transmission structure, that is the ability to transform the rotatory motion into linear motion for linear motion inerters, and the ability of magnifying the inertia of the flywheel. From this point of view, it is possible to realize other different types of inerter by using different transmission structures satisfying the above two requirements.

Although a flywheel is commonly employed in most of the current realizations of inerter, the inerter is not equivalent to the flywheel. This means that the flywheel is not always necessary to realize an inerter. In Gartner and Smith (2011), Tuluie (2010), non-flywheel inerters were introduced, where the inerter effect is achieved by rotating the fluids within some helical channels, as shown in 1.7. The theoretical inertance is represented as Gartner and Smith (2011)

$$b = \rho l \frac{A_1^2}{A_2^2}, \tag{1.5}$$

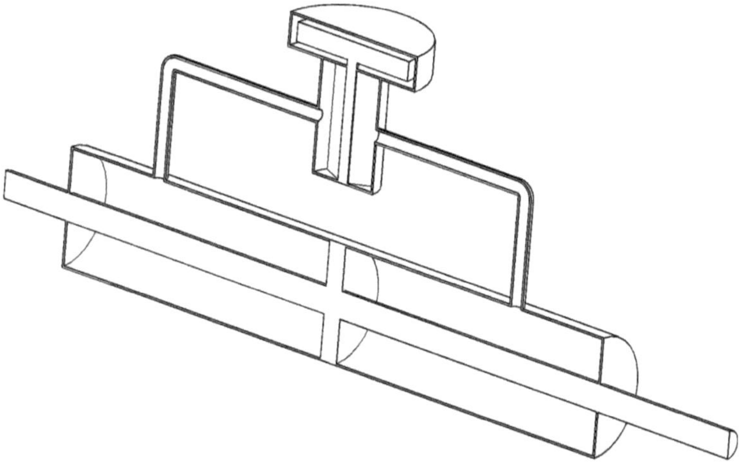

Fig. 1.6 The schematic of the flywheel-based hydraulic inerter (Wang et al. 2011)

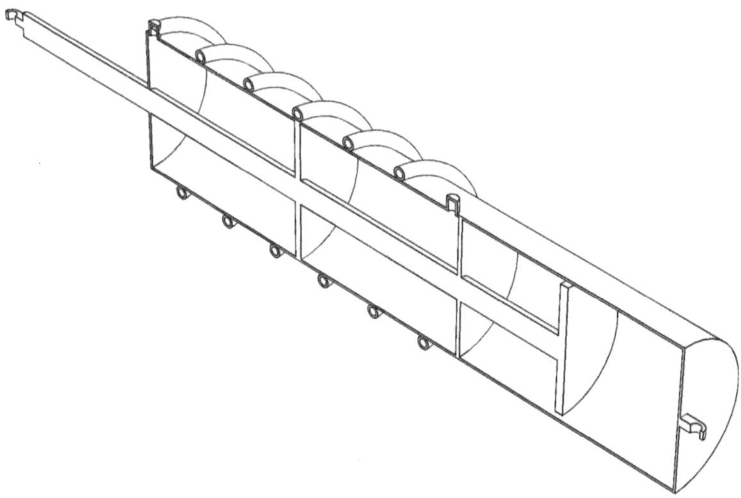

Fig. 1.7 The schematic of the non-flywheel hydraulic inerter (Gartner and Smith 2011; Tuluie 2010)

where ρ denotes the density of the fluid, l is the length of the helical channels, A_1 and A_2 denote the effective area of the piston and the helical channel, respectively.

A unique advantage of the non-flywheel inerters shown in Fig. 1.7 is that different complex inerter-spring-damper networks can be simultaneously realized, then the size of realizing such complex networks is significantly reduced (Gartner and Smith 2011). Moreover, the mass of the fluid can be smaller than the mass of the flywheel. In this way, the real physical mass of the total inerter structure can be further reduced. For example, the analysis result in Tuluie (2010) implies that 500 kg inertance can be realized by using at most 50 g fluid. This means that the magnification scale can be as large as 10000.

All the above-mentioned inerters are linear motion inerters. Actually, according to the similar definition and realization methods, rotational inerters can be similarly defined and realized, where the epicyclic gears can be used to drive a flywheel to realize a rotational inerter (Smith 2008). In this book, the linear motion inerter is mainly focused on, and for the detailed description about the rational inerter can be found in Smith (2008).

1.4 Inerter-Based Vibration Control Systems

Inerter has been successfully applied in Formula One racing cars. At the 2005 Spanish Grand Prix, Kimi Raikkonen drove the McLaren MP4-20 to victory on the first racing deployment of the inerter. To keep the technology secret from its competitors, a decoy name "J-damper" was invented for the inerter (Chen et al. 2009). Nowadays, the inerter has been employed by other Formula One teams (Chen et al. 2009).

From the mechanical control point of view, the introduction of inerter provides an extra degree of freedom compared with the traditional spring-damper mechanical systems. This ensures that the performance of the inerter-spring-damper networks is always better or at least equal to traditional spring-damper networks, as otherwise one just removes the inerter to simplify the inerter-spring-damper networks into the spring-damper ones. It has been well demonstrated that inerter can provide significant performance improvements for various mechanical systems, including vehicle suspensions, train suspensions, motorcycle steering systems, build vibration control systems, wind turbines, landing gears, bridge vibration control systems, energy harvesters, etc. The interest in passive network synthesis has also been rekindled.

In what follows, the up-to-date inerter-based vibration control systems and the main research results on inerter will be reviewed. All the results in the literature are classified from the energy consumption perspective. Therefore, three categories are obtained, namely, passive vibration control, semi-active vibration control, and active vibration control. Note that here, the passive, semi-active and active control means the control devices are passive, semi-active and active, respectively, not the overall system.

1.4.1 Passive Vibration Control with Inerters

As a passive element, one of the motivations for proposing inerter is to replace the traditional spring-damper passive vibration control devices as spring-damper-inerter ones. The methods for designing passive inerter-based vibration control systems can be generally classified into two categories, that is the fixed-structure method and the network synthesis method or "black box" method. The fixed-structure method is to investigate the performance of some specific structures. The network synthesis method is to replace the traditional spring-damper networks as a "black box" which is represented as some positive real functions. Design these positive real functions based on the required performance, and then realize the optimized positive real functions as some specific networks.

Vehicle suspensions are essential parts of a vehicle, determining the overall performance of a vehicle, which is also one of the main application fields of inerter. In Smith and Wang (2004), the performances of six inerter-based networks applied as suspension struts were numerically evaluated and compared with the traditional spring-damper strut, where it was shown that more than 10% improvements can be obtained by using inerter, in terms of the ride comfort, dynamic type load and handling performance measures. In Scheibe and Smith (2009), analytical solutions were derived for some inerter-based suspension struts based on a quarter-car model, and the performance benefits of using inerter in vehicle suspensions were analytically demonstrated. In Hu et al. (2014), multiple performance requirements including ride comfort, suspension deflection, and dynamic type load, were analytically investigated, and as a result, the direct comparison method was proposed by formulating the performance indices of complex networks as a summation of those of some simple networks and terms relating to the extra elements. In Wang and Su (2008), the nonlinearities of inerter and their influence on suspension's performance were studied. In Wang et al. (2014), two series-connected inerter-based struts were numerically and experimentally tested. In Zhang et al. (2014), an inerter-based configuration was employed to approximate the sky-hook and ground-hook dampers. In Shen et al. (2016), the dynamic vibration absorption idea was used to design an inerter-based suspension strut.

The network synthesis method was also employed in designing passive vehicle suspension. In Papageorgiou and Smith (2006), positive real synthesis using matrix inequalities was proposed. The passive suspension design problem was first formulated as a passive controller synthesis problem where the passivity constraint was described by using matrix inequalities. Then, in terms of the performances of vehicle suspensions, the H_2 and H_∞ positive real controller synthesis problem was described by using biliner matrix inequalities (BMI). After solving the BMI optimization problem, the obtained positive real controller was realized as specific mechanical networks. The effectiveness of this method was numerically and experimentally verified (Papageorgiou and Smith 2006), and this method was extended to multiple performance optimization problem in Molina-Cristobal et al. (2006). In Wang and Chan (2011), mechatronic network strut was proposed, which combined a ball-screw inerter and permanent magnet electric machinery. One of the main benefits of this mechatronic strut is system impedance can be realized through a combination of mechanical and electrical networks. As a result, higher-order system impedance can easily be realized without occupying large space. The mechatronic strut was applied to vehicle suspensions via numerical simulations and experiments (Wang and Chan 2011). In Chen et al. (2012), the performances of inerter-based strut containing one inerter and one damper, obtained by the synthesis procedure in Chen and Smith (2009), were evaluated. In Chen et al. (2015), a special class of positive real controller synthesis problem was investigated, and an efficient H_2 optimization method was proposed for passive vehicle suspensions.

Inerter has also been applied in railway vehilces. In Wang et al. (2009), the benefits of inerter for train suspensions were demonstrated via given-structure method and network synthesis based on an one-wheel train suspension model. In Wang and Liao (2010), it was demonstrated that the lateral stability can be improved by using inerter for train suspensions. In Wang et al. (2012), a full-train system was employed to verify the performance of inerter and the mechatronic strut. In Jiang et al. (2012), the inerter was demonstrated to improve the lateral and vertical ride comfort, as well as the lateral body movement. In Jiang et al. (2013), the ride quality improvement for a two-axle railway vehicle with single-stage suspension due to the employment of inerter was demonstrated.

The application of inerter in civil engineering and engineering structures constitutes one of the most concerned research areas for inerter. In Wang et al. (2010), in terms of three different building models, the performance benefits of using inerter on reducing the vibration of buildings were numerically demonstrated. In Ikago et al. (2012), a tuned viscous mass damper (TVMD) containing a ball screw mechanism and a damper was proposed applied to seismic control systems. The ball screw mechanism is actually an ball-screw inerter, where the mass of the flywheel is about 2 kg, and the inertance (in Ikago et al. (2012), it was named as apparent mass) is about 350 kg. The proposed TVMD has been implemented in a real steel structure in Japan Sugimura et al. (2012). Thereafter, the modal analysis for TVMD-based mutli-degree-of-freedom systems was conducted. In Takewaki et al. (2012), an inertial damper which was a ball-screw inerter, was propposed, and the fundamental mechanism of earthquake response reduction in building structures was studied. In Lazar et al. (2014), the performance of an inerter-based device, named tuned inerter damper (TID), was proposed to reduce vibrations in civil engineering structures subject to base excitation, where the TID is an inerter-based network composed of a series connection of a spring-damper parallel arrangement and an inerter. In Dylejko and MacGillivray (2014), a transmission absorber was proposed to tackle the internal resonance problem.

Tuned mass damper (TMD) (or dynamic vibration absorber (DVA)) is widely used in the fields of civil and mechanical engineering, which is an auxiliary mass system attached to a vibrating primary system to reduce the vibrations of the primary system. In the traditional passive TMD (or DVA), the auxiliary mass can be attached to the primary structure through a spring (proposed by Frahm in 1909 Frahm (1909)) or a spring-damper arrangement (proposed by Ormondroyd and Den Hartog in 1928 Ormondroyd and Den Hartog (1928)). In Hu and Chen (2015), the inerter-based DVA (IDVA) was proposed, and its performances were evaluated. It was demonstrated that unlike the traditional spring-damper DVA, three peaks in the frequency response were depicted for IDVA, and therefore, the optimal frequency response for IDVA was more flat than the traditional spring-damper DVA (Hu and Chen 2015). In Brzeski et al. (2015), the dynamics of TMD with additional viscous damper and inerter attached to the pendulum were investigated. In Marian and Giaralis (2014), a tuned mass-damper-inerter (TMDI) passive vibration control configuration was proposed, which connected a traditional spring-damper TMD to the neighboring mass through an inerter. The vibration suppression for mechanical cascaded systems by using TMDI was analyzed (Marian and Giaralis 2014). In Siami et al. (2018), the inerter-based isolators were applied to improve the isolation performance of the famous statue Michelangelo Buonarroti Pietǎ Rondanini.

The application of inerter in aircraft landing gears has been investigated to suppress the shimmy vibration cased by the interaction between dynamic tire behavior and landing gear structures. In Dong et al. (2015), the effect of inerters in the shimmy vibration of aircraft landing gear structures was analyzed, where it was shown the performances can be improved by using inerters, but also may yield instability problem which needs to be carefully designed. The nonlinearities in the landing gear model incorporating inerters were analyzed in Liu et al. (2015). Recently, in Li et al. (2017), the benefits of using inerter in main landing gear were further demonstrated by considering torsional-yaw, lateral, torsional-roll motions.

In terms of other applications, in Evangelou et al. (2006, 2007), the inerter-damper steering compensator was used to replace traditional one to stabilize the "weave" and "wobble" motions in two-wheel motorcycles. In Hanazawa et al. (2011), Hanazawa and Yamakita (2012), inerter was applied to flat-footed passive dynamic walkers, where it was shown that faster and more energy-efficient walking can be achieved by using ankle springs and inerters. In Graham et al. (2011), Limebeer et al. (2011), the inerter was employed to address the problem of stabilizing aeroelastic instabilities in long-span suspension bridges, and to simultaneously suppress buffeting. In Bakis et al. (2016), the improvement on deck aerodynamic performance by using inerter for a suspension bridge was investigated. In Zhao et al. (2016), wind tunnel tests were conducted to compare the theoretical and the experimental results. In Hu et al. (2015), the isolation property of several inerter-based isolators was investigated, and parameter tuning methods with respect to H_∞ and H_2 performances were proposed.

For general cascaded mechanical systems (mass-chain system), in Chen et al. (2014), the influence of inerter on natural frequencies was investigated, and it was theoretically demonstrated that the natural frequencies can be reduced by using inerters. In Yamamoto and Smith (2016), the bounded disturbance amplification for mass chains with passive interconnection was proved, and comparison between spring-damper interconnection and spring-damper-inerter interconnection was conducted. In Hu et al. (2018), the natural frequency assignment problem by using inerter was studied. It was theoretically proved that mass-chain systems with inerters may have multiple natural frequencies, where an eigenvalue of multiplicity m may occur only if $n \geq 2m - 1$. Besides, it was proved that arbitrary assignment of natural frequencies including multiplicities is not possible, but arbitrary assignment for distinct natural frequencies is always feasible and requires at most $n - 1$ inerters for an n degree-of-freedom system (Hu et al. 2018).

1.4.2 Semi-active and Active Vibration Control with Inerters

In this section, the inerter-based semi-active and active vibration control systems will be reviewed. The literatures will be classified into two categories. The first one is the passive inerter-based systems, in the sense that the inertance of the employed inerter cannot be controlled online. The second one is the semi-active inerter-based systems, namely, the inertance of the employed inerter can be controlled online.

The passive inerter-based semi-active and active vibration control systems normally consist of a inerter-based passive network and a semi-active or active actuator. For example, in Hu et al. (2012), Chen et al. (2015), the semi-active suspension system were divided into a passive part and a semi-active part, where the passive part was an inerter-based mechanical network, while the semi-active part was a semi-active damper. In Hu et al. (2018), network synthesis was employed to design the passive part by considering low-order mechanical admittances. In Zhang et al. (2012), eight inerter-based networks were combined with skyhook controlled and groundhook controlled actuators to verify the performance benefits of inerter. In Zilletti (2016), the performance of an inertial actuator consisting of a spring, a damper, an inerter and an electromechanical transducer was analyzed, where it was shown that the natural frequency reduction property of inerter can improve the stability of the overall system and hence improve the performance. In Alujeviá et al. (2018), an inerter-based source/receiving two-degree-of-freedom vibration isolation system was investigated, where the influence of inerter on the stability of the active system was analyzed, and the performance benefits of using inerter were demonstrated.

Semi-active inerter is an extension of inerter, proposed in Chen et al. (2014), defined as the inerter whose inertance can be controlled (adjusted) online (the detailed definition can be found in Chap. 5 of this book). Draw on the experience of extending passive damper to semi-active damper, it was natural to propose the semi-active inerter concept. For example, in Tsai and Huang (2011), a variable-inertia device with a magnetic planetary gearbox was proposed, where the variable inertia property can be seen as a kind of semi-active inerter. In Li et al. (2014, 2015), vehicle suspension using an adaptive inerter was studied, where the adaptive inerter was actually a semi-active inerter. However, in Li et al. (2014, 2015), no physical realization of the semi-active inerter was given.

The general physical realization of the semi-active inerter was given in Hu et al. (2017), where it was shown that there are two general solutions to realize the semi-active inerter. The first one is to adjust the transmission ratio online, and the other one is to adjust the inertia of the flywheel online. Therefore, although up to now, several realization methods for semi-active inerter have been proposed, they can always be classified into these two solutions. For example, the variable-inertia device in Tsai and Huang (2011) is realized by the first solution, that is using a magnetic planetary gearbox to adjust the transmission ratio online. Moreover, in Brzeski et al. (2015), a continuously variable transmission and gear-ratio control system was proposed to realize the stepless and accurate changes of inertance, and in Brzeski et al. (2017), experiments were conducted to verify this kind of semi-active inerters. In Lazarek et al. (2018), this semi-active inerter was applied to tuned mass damper systems by considering the parameter identification problem. To use the second solution, in Hu et al. (2017), a controllable-inertia flywheel was proposed to realize the semi-active inerter, where the performance of the proposed semi-active inerter was experimentally based on dynamic vibration absorption system.

The performance of semi-active inerter in various vibration systems have been verified. In Chen et al. (2014, 2016), the force-tracking control strategy for the semi-active inerter was proposed, and the benefits of semi-active inerter for vehicle suspensions were demonstrated. In Brzeski et al. (2017), the performance of semi-active inerter for tuned mass damper system was analyzed. In Hu et al. (2017), the semi-active inerter-based adaptive dynamic vibration absorber was proposed and its performance were verified. In Hu et al. (2017), a skyhook inerter configuration was proposed, and semi-active inerter-based realizations of the skyhook inerter configuration were studied. In Zhang et al. (2018), the realization of the skyhook inerter configuration by using hydraulic device of continuously adjustable inertance was studied.

1.5 Conclusions

In this chapter, the mechanical element inerter has been introduced, in terms of concept, physical embodiments, and the inerter-based vibration control systems. The principal motivation of proposing inerter was for mechanical network synthesis by completing the "force-current" analogy between mechanical and electrical systems. However, the inerter has brought significant impact on vibration control systems.

Due to the invention of inerter, the spring-damper-inerter mechanical networks can be systematically designed drawing the experience of electrical network synthesis. Specially, the overall system can be classified into a given part and an interconnected part to be designed. Then the given part is seen as the plant, while the part to be designed is the controller. Then, the controller is deemed as "black box", and designed from an admissible space. For example, for passive controllers, the admissible space is all the passive mechanical networks. Other similar admissible spaces could be all networks with specific number of elements etc. In this way, the parameters and the structure of the controller can be simultaneously considered (although maybe not simultaneously designed). In this way, the overall performance can be improved, and it is possible to propose new mechanical structures which could have been difficult to be recognized by the conventional methods (Smith 2002a).

Another unique property of inerter is that a very large inertance (equivalent mass) can be easily realized by using a small physical mass, which is advantageous for vibration control systems. On one hand, the mechanical structure from the "black box" method always require small masses compared with other part of the systems. For example, for vehicle suspensions, the mass of the suspensions strut should be smaller than the vehicle body and the wheel. Therefore, it is possible for the traditional spring-damper-mass mechanical impedance synthesis (Piersol and Paez 2010) to yield a large mass, but not practically feasible. On the other hand, inerter can be used to simulate mass for some circumstances, specially for civil engineering structures where large masses are beneficial but not feasible in practice.

As a mechanical concept, inerter can also be seen as a standard mechanical element, similar to the spring and damper, when modelling, analyzing and controlling mechanical systems. This means that inerter is not depended on any specific realizations, similar to the spring and damper. Actually, when modelling a mechanical system using springs, we do not always care about the spring is a steel one or a gas one. Therefore, inerter reveals the internal mechanism of some mechanical structures which at first sight are not relevant, and inerter is a standard summation of these mechanical systems.

References

Alujević, N., Čakmak, D., Wolf, H., & Jokić, M. (2018). Passive and active vibration isolation systems using inerter. *Journal of Sound and Vibration*, *418*, 163–183.

Anderson, B. D. O., & Vongpanitlerd, S. (1973). *Network analysis and synthesis*. Upper Saddle River, NJ: Prentice-Hall.

Bakis, K. N., Limebeer, D. J. N., Williams, M. S., & Graham, J. M. R. (2016). Passive aeroelastic control of a suspension bridge during erection. *Journal of Fluids and Structures*, *66*, 543–570.

Bott, R., & Duffin, R. J. (1949). Impedance synthesis without use of transformers. *Journal of Applied Physics*, *20*(8), 816.

Brzeski, P., Kapitaniak, T., & Perlikowski, P. (2015). Novel type of tuned mass damper with inerter which enables changes of inertance. *Journal of Sound and Vibration*, *349*, 56–66.

Brzeski, P., Pavlovskaia, E., Kapitaniak, T., & Perlikowski, P. (2015). The application of inerter in tuned mass absorber. *International Journal of Non-Linear Mechanics*, *70*, 20–29.

Brzeski, P., Lazarek, M., & Perlikowski, P. (2017). Experimental study of the novel tuned mass damper with inerter which enables changes of inertance. *Journal of Sound and Vibration*, *404*, 47–57.

Chen, M. Z. Q., & Smith, M. C. (2009). Restricted complexity network realizations for passive mechanical control. *IEEE Transactions on Automatic Control*, *54*(10), 2290–2301.

Chen, M. Z. Q., & Smith, M. C. (2009). A note on tests for positive-real functions. *IEEE Transactions on Automatic Control*, *54*(2), 390–393.

Chen, M. Z. Q., Papageorgiou, C., Scheibe, F., Wang, F. C., & Smith, M. C. (2009). The missing mechanical circuit element. *IEEE Circuits and Systems Magazine*, *9*(1), 10–26.

Chen, M. Z. Q., Hu, Y., & Du, B. (2012). Suspension performance with one damper and one inerter. In *Proceedings of the 24th Chinese Control and Decision Conference*, Taiyuan, China (pp. 3551–3556).

Chen, M. Z. Q., Hu, Y., Huang, L., & Chen, G. (2014). Influence of inerter on natural frequencies of vibration systems. *Journal of Sound and Vibration*, *333*(7), 1874–87.

Chen, M. Z. Q., Hu, Y., Li, C., & Chen, G. (2014). Semi-active suspension with semi-active inerter and semi-active damper. *IFAC Proceedings Volumes*, *47*(3), 11225–11230.

Chen, M. Z. Q., Hu, Y., & Wang, F.-C. (2015). Passive mechanical control with a special class of positive real controllers: Application to passive vehicle suspensions. *Journal of Dynamic Systems, Measurement, and Control*, *137*(12), 121013.

Chen, M. Z. Q., Hu, Y., Li, C., & Chen, G. (2015). Performance benefits of using inerter in semiactive suspensions. *IEEE Transactions on Control Systems Technology*, *23*(4), 1571–1577.

Chen, M. Z. Q., Hu, Y., Li, C., & Chen, G. (2016). Application of semi-active inerter in semi-active suspensions via force tracking. *Journal of Vibration and Acoustics*, *138*(4), 041014.

Chen, M. Z. Q., Wang, K., Zou, Y., & Lam, J. (2013). Realization of a special class of admittances with one damper and one inerter for mechanical control. *IEEE Transactions on Automatic Control*, *58*(7), 1841–1846.

Chen, M. Z. Q., Wang, K., Zou, Y., & Chen, G. (2015). Realization of three-port spring networks with inerter for effective mechanical control. *IEEE Transactions on Automatic Control*, *60*(10), 2722–2727.

Dong, X., Liu, Y., & Chen, M. Z. Q. (2015). Application of inerter to aircraft landing gear suspension. In *Proceedings of the 34th Chinese Control Conference*, July 28–30, Hangzhou, China (pp. 2066–2071).

Dylejko, P. G., & MacGillivray, I. R. (2014). On the concept of a transmission absorber to suppress internal resonance. *Journal of Sound and Vibration*, *333*, 2719–2734.

Evangelou, S., Limebeer, D. J. N., Sharp, R. S., & Smith, M. C. (2006). Control of motorcycle steering instabilities. *IEEE Control Systems Magazine*, *26*(5), 78–88.

Evangelou, S., Limebeer, D. J. N., Sharp, R. S., & Smith, M. C. (2007). Steering compensation for high-performance motorcycles. *Journal of Applied Mechanics*, *74*(2), 332–346.

Frahm, H. (1909). *Device for damping vibrations of bodies*. U.S. Patent, No. 989958. 30.

Gartner, B. J., & Smith, M. C. (2011). Damper and inertial hydraulic device. U.S. Patent 13/577, 234.

Graham, J. M. R., Limebeer, D. J. N., & Zhao, X. (2011). Aeroelastic control of long-span suspension bridges. *Journal of Applied Mechanics*, *78*(4), 041018.

Hanazawa, Y., Suda, H., & Yamakita, M. (2011). Analysis and experiment of flat-footed passive dynamic walker with ankle inerter. In *Proceedings of the 2011 IEEE International Conference on Robotics and Biomimetics*, December 7–11, Phuket, Thailand (pp. 86–91).

Hanazawa, Y., & Yamakita, M. (2012). High-efficient biped walking based on flat-footed passive dynamic walking with mechanical impedance at ankles. *Journal of Robotics and Mechatronics*, *24*(3), 498–506.

Hu, Y., Li, C., & Chen, M. Z. Q. (2012). Optimal control for semi-active suspension with inerter. In *Proceedings of the 31st Chinese Control Conference*, Hefei, China (pp. 2301–2306).

Hu, Y., & Chen, M. Z. Q. (2015). Performance evaluation for inerter-based dynamic vibration absorbers. *International Journal of Mechanical Sciences*, *99*, 297–307.

Hu, Y., Wang, K., Chen, Y., & Chen, M. Z. Q. (2018). Inerter-based semi-active suspensions with low-order mechanical admittance via network synthesis. *Transactions of the Institute of Measurement and Control*. 0142331217744852.

Hu, Y., Chen, M. Z. Q., & Shu, Z. (2014). Passive vehicle suspensions employing inerters with multiple performance requirements. *Journal of Sound and Vibration*, *333*, 2212–2225.

Hu, Y., Chen, M. Z. Q., Shu, Z., & Huang, L. (2015). Analysis and optimisation for inerter-based isolators via fixed-point theory and algebraic solution. *Journal of Sound and Vibration*, *346*, 17–36.

Hu, Y., Chen, M. Z. Q., & Smith, M. C. (2018). Natural frequency assignment for mass-chain systems with inerters. *Mechanical Systems and Signal Processing*, *108*, 126–139.

Hu, Y., Chen, M. Z. Q., & Sun, Y. (2017). Comfort-oriented vehicle suspension design with skyhook inerter configuration. *Journal of Sound and Vibration*, *405*, 34–47.

Hu, Y., Chen, M. Z. Q., Xu, S., & Liu, Y. (2017). Semiactive inerter and its application in adaptive tuned vibration absorbers. *IEEE Transactions on Control Systems Technology*, *25*(1), 294–300.

Ikago, K., Saito, K., & Inoue, N. (2012). Seismic control of SDOF structure using tuned viscous mass damper. *Earthquake Engineering and Structural Dynamics*, *41*, 453–474.

Ikago, K., Sugimura, Y., Saito, K., & Inoue, K. (2012). Modal response characteristics of a multiple-degree-of-freedom structure incorporated with tuned viscous mass damper. *Journal of Asian Architecture and Building Engineering*, *11*, 375–382.

Jiang, J. Z., Matamoros-Sanchez, A. Z., Zolotas, A., Goodall, R. M., & Smith, M. C. (2013). Passive suspensions for ride quality improvement of two-axle railway vehicles. *Proceedings of the Institution of Mechanical Engineers, Part F: Journal of Rail and Rapid Transit*. 0954409713511592.

Jiang, J. Z., Matamoros-Sanchez, A. Z., Goodall, R. M., & Smith, M. C. (2012). Passive suspensions incorporating inerters for railway vehicles. *Vehicle System Dynamics, 50*, 263–276.

Lazarek, M., Brzeski, P., & Perlikowski, P. (2018). Design and identification of parameters of tuned mass damper with inerter which enables changes of inertance. *Mechanism and Machine Theory, 119*, 161–173.

Lazar, I. F., Neild, S. A., & Wagg, D. J. (2014). Using an inerter-based device for structural vibration suppression. *Earthquake Engineering and Structure Dynamics, 43*(8), 1129–1147.

Li, P., Lam, J., & Cheung, K. C. (2014). Investigation on semi-active control of vehicle suspension using adaptive inerter. In *The 21st International Congress on Sound and Vibration*, Beijing, China.

Li, Y., Howcroft, C., Neild, S. A., & Jiang, J. Z. (2017). Using continuation analysis to identify shimmy-suppression devices for an aircraft main landing gear. *Journal of Sound and Vibration, 408*, 234–251.

Li, Y., Jiang, J. Z., & Neild, S. (2017). Inerter-based configurations for main-landing-gear shimmy suppression. *Journal of Aircraft, 54*(2), 684–693.

Li, Y., Jiang, J. Z., Neild, S. A., & Wang, H. (2017). Optimal inerter-based shock-strut configurations for landing-gear touchdown performance. *Journal of Aircraft, 54*(5), 1901–1909.

Li, P., Lam, J., & Cheung, K. C. (2015). Control of vehicle suspension using an adaptive inerter. *Proceedings of the Institution of Mechanical Engineers, Part D: Journal of Automobile Engineering, 229*(14), 1934–1943.

Limebeer, D. J. N., Graham, J. M. R., & Zhao, X. (2011). Buffet suppression in long-span suspension bridges. *Annual Reviews in Control, 35*(2), 235–246.

Liu, Y., Chen, M. Z. Q., & Tian, Y. (2015). Nonlinearities in landing gear model incorporating inerter. In *Proceeding of the 2015 IEEE International Conference on Information and Automation*, Lijiang, China (pp. 696–701)

Marian, L., & Giaralis, A. (2014). Optimal design of a novel tuned mass-damper-inerter (TMDI) passive vibration control configuration for stochastically support-excited structural systems. *Probabilistic Engineering Mechanics, 38*, 156–164.

Molina-Cristobal, A., Papageorgiou, C., Parks, G. T., Smith, M. C., & Clarkson, P. J. (2006). Multi-objective controller design: Evolutionary algorithms and bilinear matrix inequalities for a passive suspension. In *Proceedings of the 13th IFAC Workshop on Control Applications of Optimization*, Cachan-Paris, France (pp. 386–391).

Ormondroyd, J., & Den Hartog, J. P. (1928). The theory of the dynamic vibration absorber. *ASME Journal of Applied Mechanics, 50*, 9–22.

Papageorgiou, C., & Smith, M. C. (2005). Laboratory experimental testing of inerters. In *Proceedings of the 44th IEEE Conference on Decision and Control, and the European Control Conference 2005*, Seville, Spain, December 12–15 (pp. 3351–3356).

Papageorgiou, C., & Smith, M. C. (2006). Positive real synthesis using matrix inequalities for mechanical networks: Application to vehicle suspension. *IEEE Transactions on Control Systems Technology, 14*(3), 423–435.

Piersol, A. G., & Paez, T. L. (2010). *Harris' shock and vibration handbook* (6th ed.). New York: McGraw-Hill.

Scheibe, F., & Smith, M. C. (2009). Analytical solutions for optimal ride comfort and tyre grip for passive vehicle suspensions. *Vehicle System Dynamics, 47*(10), 1229–1252.

Shen, Y., Chen, L., Yang, X., Shi, D., & Yang, J. (2016). Improved design of dynamic vibration absorber by using the inerter and its application in vehicle suspension. *Journal of Sound and Vibration, 361*, 148–158.

Siami, A., Karimi, H. R., Cigada, A., Zappa, E., & Sabbioni, E. (2018). Parameter optimization of an inerter-based isolator for passive vibration control of Michelangelos Rondanini Piet. *Mechanical Systems and Signal Processing, 98*, 667–683.

Smith, M. C. (2002a). Synthesis of mechanical networks: The inerter. *IEEE Transactions on Automatic Control*, *47*(1), 1648–1662.
Smith, M. C. (2002b). Force-controlling mechanical device. U.S. Patent 7,316,303 B2.
Smith, M. C. (2003). The inerter concept and its application. Plenary Lecture, Society of Instrument and Control Engineers (SICE) Annual Conference Fukui, Japan 4 August 2003.
Smith, M. C. (2008). Force-controlling mechanical device. U.S. Patent 7/316 303.
Smith, M. C. (2011). Vehicle dynamics, engineering thought-experiments and Formula One racing. William Mong Distinguished Lecture, The University of Hong Kong, 13 January 2011
Smith, M. C., & Wang, F.-C. (2004). Performance benefits in passive vehicle suspensions employing inerters. *Vehicle System Dynamics*, *42*(4), 235–257.
Sugimura, Y., Goto, W., Tanizawa, H., Saito, K., & Nimomiya, T. (2012). Response control effect of steel building structure using tuned viscous mass damper. In *Proceedings of the 15th World Conference on Earthquake Engineering*, Lisbon, Portugal.
Takewaki, I., Murakami, S., Yoshitomi, S., & Tsuji, M. (2012). Fundamental mechanism of earthquake response resuction in buildind structures with inertial dampers. *Journal of Structural Control and Health Monitoring*, *19*, 590–608.
Tsai, M. C., & Huang, C. C. (2011). Development of a variable-inertia device with a magnetic planetary gearbox. *IEEE/ASME Transactions on mechatronics*, *16*(6), 1120–1128.
Tuluie, R. (2010). Fluid Inerter. U.S. Patent 13/575, 017.
Wang, K., Chen, M. Z. Q., Li, C., & Chen, G. (2018). Passive controller realization of a biquadratic impedance with double poles and zeros as a seven-element series-parallel network for effective mechanical control. *IEEE Transactions on Automatic Control*. https://doi.org/10.1109/TAC.2018.2794820 (in press).
Wang, F.-C., Hsu, M.-S., Su, W.-J., & Lin, T. C. (2009). Screw type inerter mechanism. U.S. Patent 2009/0108510 A1.
Wang, R., Meng, X., Shi, D., Zhang, X., Chen, Y., & Chen, L. (2014). Design and test of vehicle suspension system with inerters. *Proceedings of the Institution of Mechanical Engineers, Part C: Journal of Mechanical Engineering Science*, 1–6.
Wang, F.-C., & Chan, H.-A. (2011). Vehicle suspensions with a mechatronic network strut. *Vehicle System Dynamics*, *49*(5), 811–830.
Wang, F.-C., Hong, M. F., & Chen, C. W. (2010). Building suspensions with inerters. *Proceedings of the IMechE, Part C: Journal of Mechanical Engineering Science*, *224*(8), 1605–1616.
Wang, F.-C., Hong, M. F., & Lin, T. C. (2011). Designing and testing a hydraulic inerter. *Proceedings of the Institution of Mechanical Engineers, Part C: Journal of Mechanical Engineering Science*, *225*(1), 66–72.
Wang, F.-C., Hsieh, M.-R., & Chen, H.-J. (2012). Stability and performance analysis of a full-train system with inerters. *Vehicle System Dynamics*, *50*(4), 545–571.
Wang, F.-C., & Liao, M.-K. (2010). The lateral stability of train suspension systems employing inerters. *Vehicle System Dynamics*, *8*(5), 619–643.
Wang, F.-C., Liao, M. K., Liao, B. H., Su, W. J., & Chan, H. A. (2009). The performance improvements of train suspension systems with mechanical networks employing inerters. *Vehicle System Dynamics*, *47*(7), 805–830.
Wang, F.-C., & Su, W.-J. (2008). Impact of inerter nonlinearities on vehicle suspension control. *Vehicle System Dynamics*, *46*(7), 575–595.
Yamamoto, K., & Smith, M. C. (2016). Bounded disturbance amplification for mass chains with passive interconnection. *IEEE Transactions on Automatic Control*, *61*(6), 1565–1574.
Zhang, X. L., Zhang, T., Nie, J., & Chen, L. (2018). A semiactive skyhook-inertance control strategy based on continuously adjustable inerter. *Shock and Vibration*, 6828621.
Zhang, X. J., Ahmadian, M., & Guo, K. H. (2012). On the benefits of semi-active suspensions with inerters. *Shock and Vibration*, *19*(3), 257–272.
Zhang, X. L., Liu, J. J., Nie, J. M., & Chen, L. (2014). Design principle and method of a passive hybrid damping suspension system. *Applied Mechanics and Materials*, *635–637*, 1232–1240.

Zhao, X., Gouder, K., Graham, J. M. R., & Limebeer, D. J. (2016). Buffet loading, dynamic response and aerodynamic control of a suspension bridge in a turbulent wind. *Journal of Fluids and Structures*, *62*, 384–412.

Zilletti, M. (2016). Feedback control unit with an inerter proof-mass electrodynamic actuator. *Journal of Sound and Vibration*, *369*, 16–28.

Chapter 2
Analysis for Inerter-Based Vibration System

Abstract This chapter investigates the influence of inerter on the natural frequencies of vibration systems. First of all, the natural frequencies of a single-degree-of-freedom (SDOF) system and a two-degree-of-freedom (TDOF) system are derived algebraically and the fact that inerter can reduce the natural frequencies of these systems is demonstrated. Then, to further investigate the influence of inerter in a general vibration system, a multi-degree-of-freedom system (MDOF) is considered. Sensitivity analysis is performed on the natural frequencies and mode shapes to demonstrate that the natural frequencies of the MDOF system can always be reduced by increasing the inertance of any inerter. The condition for a general MDOF system of which the natural frequencies can be reduced by an inerter is also derived. Finally, the influence of inerter position on the natural frequencies is investigated and the efficiency of inerter in reducing the largest natural frequencies is verified by simulating a six-degree-of-freedom system, where a reduction of more than 47% is obtained by employing only five inerters.

Keywords Natural frequency · Single-degree-of-freedom system · Two-degree-of-freedom system · Multi-degree-of-freedom system · Sensitivity analysis

2.1 Introduction

Inerter has been applied in various mechanical systems. However, among these applications, inerter always appears in some mechanical networks which possess more complex structures than the conventional networks consisting of only springs and dampers. The networks with inerters will surely be better than or at least equal to the conventional networks consisting of only springs and dampers as they can always reduce to the conventional ones when the values of element coefficients (spring stiffness, damping coefficient, or inertance) become zero or infinity (Chen et al. 2012). It is true that inerter can provide extra flexibility in structure, but the basic functionality of inerter in vibration systems has not yet been clearly understood and demonstrated.

It is well known that in a vibration system, spring can store energy, provide static support, and determine the natural frequencies, while viscous damper can dissipate

M. Z. Q. Chen and Y. Hu, *Inerter and Its Application in Vibration Control Systems*, https://doi.org/10.1007/978-981-10-7089-1_2

energy, limit the amplitude of oscillation at resonance, and slightly decrease the natural frequencies if the damping is small (Tomson 1993). As shown in Smith (2002), inerter can store energy. However, for the other inherent properties of vibration systems such as natural frequencies, the influence of inerter has not been investigated before.

The objective of this chapter is to study the fundamental influence of inerter on the natural frequencies of vibration systems. The fact that inerter can reduce the natural frequencies of vibration systems is theoretically demonstrated in this chapter and the question that how to efficiently use inerter to reduce the natural frequencies is also addressed.

2.2 Preliminary

It is well known that all systems containing mass and elasticity are capable of free vibration, that is, the vibration occurring without external excitation (Tomson 1993). Natural frequency of vibration is of primary interest for such systems. For a single-degree-of-freedom spring–mass system shown in Fig. 2.1, the motion of equation can be written as

$$m\ddot{x} + c\dot{x} + kx = 0.$$

In another form,

$$\ddot{x} + 2\zeta\omega_n\dot{x} + \omega_n^2 x = 0, \tag{2.1}$$

where

$$\omega_n = \sqrt{\frac{k}{m}}, \ \zeta = \frac{c}{2\sqrt{mk}}.$$

Here, ω_n is called *natural frequency* and ζ is the mode damping coefficient.

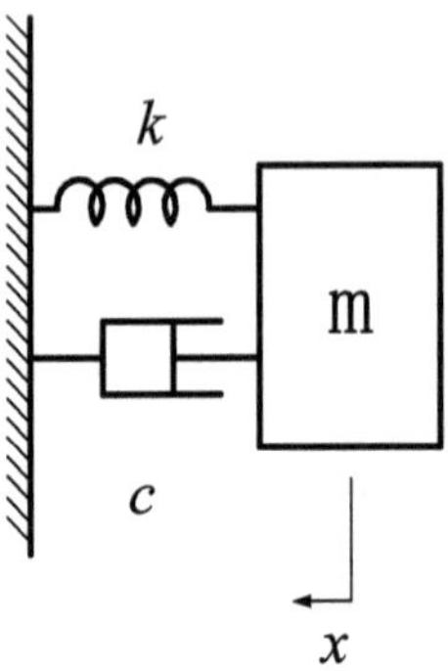

Fig. 2.1 A single-degree-of-freedom spring–mass system

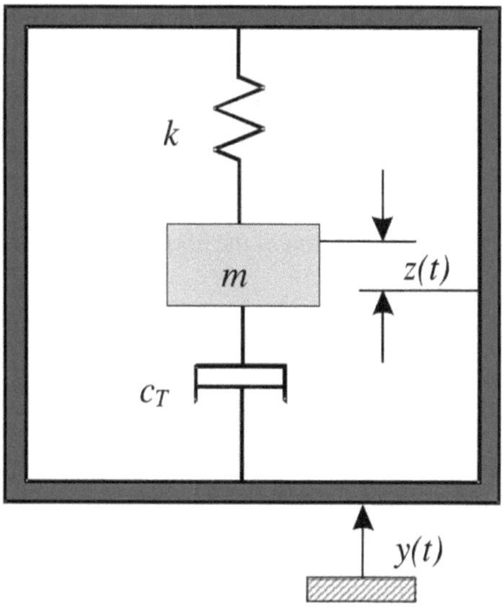

Fig. 2.2 Model of a vibration-based self-powered system

Since the influence of damping on natural frequencies is well known, only the undamped conservative systems are considered for simplicity. For the undamped system, i.e., $\zeta = 0$, the solution of (2.1) is

$$x(t) = \frac{\dot{x}(0)}{\omega_n} \sin \omega_n t + x(0) \cos \omega_n t,$$

where $\dot{x}(0)$ and $x(0)$ are the initial velocity and displacement. This implies that the system harmonically vibrates at the natural frequency.

For forced vibration cases, when the frequency of the excitation is equal to one of the natural frequencies, there may occur a phenomenon known as *resonance*, which may lead to excessive deflections and failure (Tse et al. 1979). In practice, it is always desirable to adjust the natural frequencies of a vibration system to avoid or induce resonance where appropriate. For example, for vibration-based self-powered systems (Beeby et al. 2006) (as shown in Fig. 2.2), the natural frequency of an embedded spring–mass system should be consistent with the environment to obtain maximum vibration power by utilizing resonance, while for the engine mounting systems (Yu et al. 2001), the natural frequency should be below the engine disturbance frequency of the engine idle speed to avoid excitation of mounting system resonance.

The traditional methods to reduce the natural frequencies of an elastic system are either decreasing the elastic stiffness or increasing the mass of the vibration system. However, this may be problematic; for example, the stiffness values of an engine mount that are too low will lead to large static and quasi-static engine displacements and damage of some engine components (Yu et al. 2001). It will be shown below that other than these two methods, a parallel-connected inerter can also effectively reduce natural frequencies.

2.3 Single-Degree-of-Freedom System

A SDOF system with an inerter is shown in Fig. 2.3. The equation of motion for free vibration of this system is

$$(m + b)\ddot{x} + kx = 0. \tag{2.2}$$

Transformation of the above equation into the standard form for vibration analysis yields

$$\ddot{x} + \omega_n^2 x = 0,$$

where $\omega_n = \sqrt{\frac{k}{m+b}}$ is called the natural frequency of the undamped system.

Proposition 1 *The natural frequency ω_n of an SDOF system is a decreasing function of the inertance b. Thus, inerter can reduce the natural frequency of an SDOF system.*

Remark 2.1 Note that in Smith (2002), one application of inerter is to simulate the mass by connecting a terminal of an inerter to the mechanical ground. Observing (2.2), one concludes that the inerter with one terminal connected to ground can effectively enlarge the mass which is connected at the other terminal.

2.4 Two-Degree-of-Freedom System

To investigate the general influence of inerter on the natural frequencies of a vibration system, a TDOF system, shown in Fig. 2.4, is investigated in this section.

The equations of motion for free vibration of this system are

$$\begin{aligned} m_1\ddot{x}_1 + k_1(x_1 - x_2) + b_1(\ddot{x}_1 - \ddot{x}_2) &= 0, \\ m_2\ddot{x}_2 - k_1(x_1 - x_2) - b_1(\ddot{x}_1 - \ddot{x}_2) + k_2x_2 + b_2\ddot{x}_2 &= 0, \end{aligned}$$

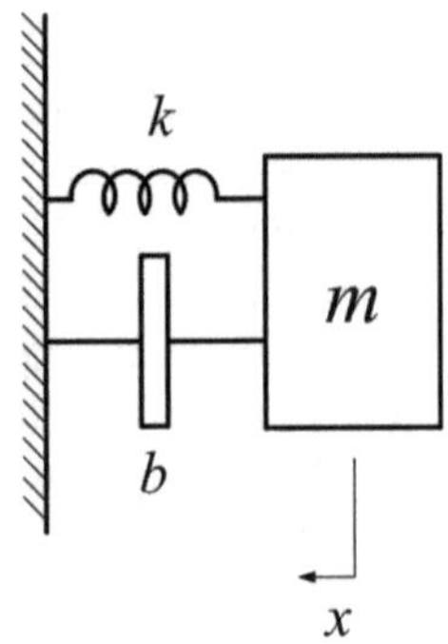

Fig. 2.3 SDOF system with an inerter

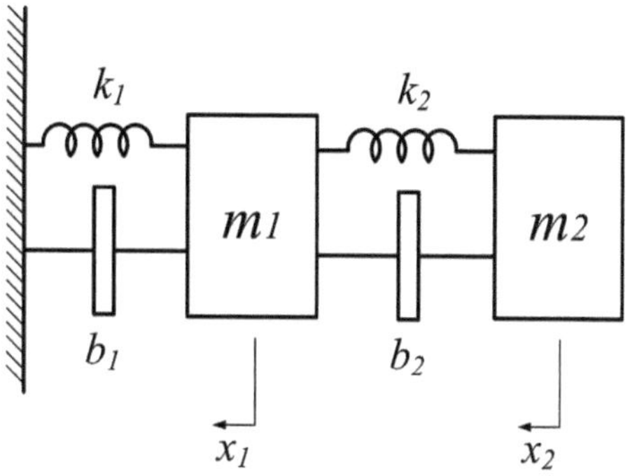

Fig. 2.4 TDOF system with two inerters

or, in a compact form,

$$\mathbf{M}\ddot{\mathbf{x}} + \mathbf{K}\mathbf{x} = 0,$$

where **M** is called the inertia matrix and **K** is the stiffness matrix (Tse et al. 1979), and

$$\mathbf{M} = \begin{bmatrix} m_1 + b_1 & -b_1 \\ -b_1 & m_2 + b_1 + b_2 \end{bmatrix}, \quad \mathbf{K} = \begin{bmatrix} k_1 & -k_1 \\ -k_1 & k_1 + k_2 \end{bmatrix}.$$

Note that the inertances b_1 and b_2 only exist in the inertia matrix **M**, but the positions of b_1 and b_2 are different as b_1 exists in all the elements of **M** while b_2 only appears in the last element of **M**. Since one terminal of b_2 is connected to the ground, b_2 effectively enlarges the mass m_2, which is consistent with the conclusion made in Remark 2.1.

The two natural frequencies can be obtained by solving the characteristic equation (Tse et al. 1979)

$$\begin{aligned} \Delta(\omega) &= \left|\mathbf{K} - \mathbf{M}\omega^2\right| \\ &= (m_1m_2 + m_1(b_1 + b_2) + m_2b_1 + b_1b_2)\omega^4 - ((m_1 + m_2)k_1 + m_1k_2 + \\ &\quad k_1b_2 + b_1k_2)\omega^2 + k_1k_2 = 0, \end{aligned} \tag{2.3}$$

which yields

$$\omega_{n1} = \sqrt{\frac{k_1k_2(f_1 + f_2 - \sqrt{(f_1 - f_2)^2 + 4d_0})}{2(f_1f_2 - d_0)}}, \tag{2.4}$$

$$\omega_{n2} = \sqrt{\frac{k_1k_2(f_1 + f_2 + \sqrt{(f_1 - f_2)^2 + 4d_0})}{2(f_1f_2 - d_0)}}, \tag{2.5}$$

where $f_1 = (m_1 + m_2 + b_2)k_1$, $f_2 = (m_1 + b_1)k_2$, and $d_0 = k_1k_2m_1^2$.

Proposition 2 *For a TDOF system with two inerters, both natural frequencies* ω_{n1} *and* ω_{n2} *are decreasing functions of the inertances* b_1 *and* b_2.

Proof The monotonicity of ω_{n1} and ω_{n2} can be proven by checking the signs of the first-order derivatives of ω_{n1}^2 and ω_{n2}^2 in terms of f_1 and f_2, respectively.

$$\frac{\partial \omega_{n1}^2}{\partial f_1} = -\frac{k_1 k_2 (q_1 - q_2)}{2(d_0 - f_1 f_2)^2 \sqrt{(f_1 - f_2)^2 + 4d_0}},$$
$$\frac{\partial \omega_{n2}^2}{\partial f_1} = -\frac{k_1 k_2 (q_1 + q_2)}{2(d_0 - f_1 f_2)^2 \sqrt{(f_1 - f_2)^2 + 4d_0}},$$

where $q_1 = (d_0 + f_2^2)\sqrt{(f_1 - f_2)^2 + 4d_0}$ and $q_2 = f_1(d_0 - f_2^2) + 3f_2 d_0 + f_2^3$.

Note that $q_1 > 0$ and

$$q_1^2 - q_2^2 = 4d_0 f_2^2 (f_1 - d_0/f_2)^2,$$

so one obtains $|q_1| > |q_2|$, which implies $\frac{\partial \omega_{n1}^2}{\partial f_1} < 0$ and $\frac{\partial \omega_{n2}^2}{\partial f_1} < 0$, that is, both ω_{n1} and ω_{n2} are decreasing functions of inertance b_2.

Similarly,

$$\frac{\partial \omega_{n1}^2}{\partial f_2} = -\frac{k_1 k_2 (q_3 - q_4)}{2(d_0 - f_1 f_2)^2 \sqrt{(f_1 - f_2)^2 + 4d_0}},$$
$$\frac{\partial \omega_{n2}^2}{\partial f_2} = -\frac{k_1 k_2 (q_3 + q_4)}{2(d_0 - f_1 f_2)^2 \sqrt{(f_1 - f_2)^2 + 4d_0}},$$

where $q_3 = (d_0 + f_1^2)\sqrt{(f_1 - f_2)^2 + 4d_0}$ and $q_4 = f_2(d_0 - f_1^2) + 3f_1 d_0 + f_1^3$.

Since $q_3 > 0$ and $q_3^2 - q_4^2 = 4d_0 f_1^2 (f_2 - d_0/f_1)^2 > 0$, one has $|q_3| > |q_4|$, $\frac{\partial \omega_{n1}^2}{\partial f_2} < 0$, and $\frac{\partial \omega_{n2}^2}{\partial f_2} < 0$, that is, both ω_{n1} and ω_{n2} are decreasing functions of inertance b_1. □

2.5 Multi-degree-of-Freedom System

From the previous two sections, one sees that inerter can reduce the natural frequencies of both SDOF and TDOF systems. To find out whether this holds for any vibration system, a general MDOF system, shown in Fig. 2.5, is investigated in this section.

The equations of motion of the MDOF system shown in Fig. 2.5 are

$$\mathbf{M}\ddot{\mathbf{x}} + \mathbf{K}\mathbf{x} = 0,$$

where $\mathbf{x} = [x_1, x_2, \ldots, x_n]^T$, and

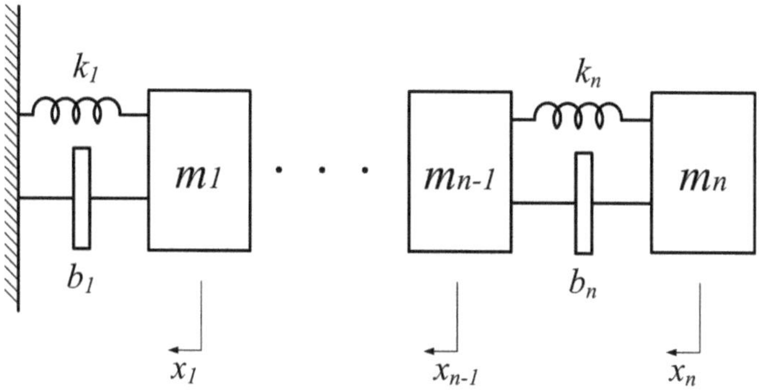

Fig. 2.5 MDOF system with inerters

$$\mathbf{M} = \begin{bmatrix} m_1 + b_1 & -b_1 & & & \\ -b_1 & m_2 + b_1 + b_2 & -b_2 & & \\ & \ddots & \ddots & \ddots & \\ & & & -b_{n-1} & m_n + b_{n-1} + b_n \end{bmatrix},$$

$$\mathbf{K} = \begin{bmatrix} k_1 & -k_1 & & \\ -k_1 & k_1 + k_2 & -k_2 & \\ \ddots & \ddots & \ddots & \\ & & -k_{n-1} & k_{n-1} + k_n \end{bmatrix}.$$

It is well known that the free vibration of the MDOF system can be described by the eigenvalue problem as follows (Tomson 1993; Zhao and DeWolf 1999)

$$(\mathbf{K} - \mathbf{M}\lambda_j)\boldsymbol{\varphi}_j = \mathbf{0}, \tag{2.6}$$

where $j = 1, \ldots, n$, $\omega_{ni} = \sqrt{\lambda_j}$ are the natural frequencies of this system, and $\boldsymbol{\varphi}_j$ is the jth mode shape corresponding to natural frequency ω_{nj} and is normalized to be unit-mass mode shapes, i.e., $\boldsymbol{\varphi}_j{}^T\mathbf{M}\boldsymbol{\varphi}_j = 1$.

Sensitivity analysis is performed on the eigenvalues and eigenvectors with respect to each inertance and the following proposition is derived.

Proposition 3 *Consider the MDOF system shown in Fig. 2.5. For an arbitrary eigenvalue λ_j, $j = 1, \ldots, n$, and an arbitrary inertance b_i, $i = 1, \ldots, n$, the following equations hold:*

$$\frac{\partial \lambda_j}{\partial b_i} = -\lambda_j \Phi_{ij}, \tag{2.7}$$

$$\frac{\partial \Phi_{ij}}{\partial b_i} = 2\Phi_{ij}\left(-\frac{1}{2}\Phi_{ij} + \sum_{l=1, l\neq j}^{n} \frac{\lambda_j}{\lambda_l - \lambda_j}\Phi_{il}\right), \tag{2.8}$$

$$\frac{\partial^2 \lambda_j}{\partial b_i^2} = 2\lambda_j \Phi_{ij} \left(\Phi_{ij} - \sum_{l=1, l \neq j}^{n} \frac{\lambda_j}{\lambda_l - \lambda_j} \Phi_{il} \right), \tag{2.9}$$

where Φ_{ij}, $j = 1, \ldots, n$, is defined as

$$\Phi_{ij} = {\boldsymbol{\varphi}_j}^T \frac{\partial \mathbf{M}}{\partial b_i} \boldsymbol{\varphi}_j = \begin{cases} \left(\boldsymbol{\varphi}_j^{(i)} - \boldsymbol{\varphi}_j^{(i+1)} \right)^2, & i \neq n \\ \left(\boldsymbol{\varphi}_j^{(n)} \right)^2, & i = n \end{cases}$$

Proof The proof is inspired by the sensitivity analysis on natural frequencies (eigenvalues) and model shapes (eigenvectors) with respect to structure parameters in Zhao and DeWolf (1999), Lin and Parker (1999), Lee and Kim (1999).

Sensitivity analysis on natural frequencies:
Considering the influence of the ith inertance b_i on the jth natural frequency ω_{nj}, the derivative of (2.6) with respect to b_i is

$$\left(\frac{\partial \mathbf{K}}{\partial b_i} - \frac{\partial \lambda_j}{\partial b_i} \mathbf{M} - \lambda_j \frac{\partial \mathbf{M}}{\partial b_i} \right) \boldsymbol{\varphi}_j + (\mathbf{K} - \lambda_j \mathbf{M}) \frac{\partial \boldsymbol{\varphi}_j}{\partial b_i} = 0. \tag{2.10}$$

Premultiplying both sides of (2.10) by ${\boldsymbol{\varphi}_j}^T$ and considering the relations that $\frac{\partial \mathbf{K}}{\partial b_i} = 0$ ($\mathbf{K}$ is independent of b_i), ${\boldsymbol{\varphi}_j}^T(\mathbf{K} - \lambda_j \mathbf{M}) = 0$, and ${\boldsymbol{\varphi}_j}^T \mathbf{M} \boldsymbol{\varphi}_j = 1$, one obtains

$$\frac{\partial \lambda_j}{\partial b_i} = -\lambda_j {\boldsymbol{\varphi}_j}^T \frac{\partial \mathbf{M}}{\partial b_i} \boldsymbol{\varphi}_j = 0. \tag{2.11}$$

Note that

$$\frac{\partial \mathbf{M}}{\partial b_i} = \begin{cases} \begin{bmatrix} 0 & & & & & \\ & \ddots & & & & \\ & & 1 & -1 & & \\ & & -1 & 1 & & \\ & & & & \ddots & \\ & & & & & 0 \end{bmatrix}, & i \neq n \\ \begin{bmatrix} 0 & & & \\ & \ddots & & \\ & & 0 & \\ & & & 1 \end{bmatrix}, & i = n \end{cases} \tag{2.12}$$

where the nonzero elements for the case $i \neq n$ locate on the ith, $i+1$th rows and ith, $i+1$th columns.

Thus, one obtains

$$\frac{\partial \lambda_j}{\partial b_i} = \begin{cases} -\lambda_j \left(\boldsymbol{\varphi}_j{}^{(i)} - \boldsymbol{\varphi}_j{}^{(i+1)}\right)^2, & i \neq n \\ -\lambda_j \left(\boldsymbol{\varphi}_j{}^{(n)}\right)^2, & i = n \end{cases} \tag{2.13}$$

where $\boldsymbol{\varphi}_j{}^{(i)}$, $i = 1, \ldots, n$, denotes the ith element of $\boldsymbol{\varphi}_j$.

Denoting

$$\Phi_{ij} = \boldsymbol{\varphi}_j{}^T \frac{\partial \mathbf{M}}{\partial b_i} \boldsymbol{\varphi}_j = \begin{cases} \left(\boldsymbol{\varphi}_j^{(i)} - \boldsymbol{\varphi}_j^{(i+1)}\right)^2, & i \neq n \\ \left(\boldsymbol{\varphi}_j^{(n)}\right)^2, & i = n \end{cases}$$

where $j = 1, \ldots, n$, one obtains (2.7). □

It is clearly shown in (2.7) that

$$\frac{\partial \lambda_j}{\partial b_i} \leq 0,$$

and the equality is achieved if $\boldsymbol{\varphi}_j{}^{(i)} = \boldsymbol{\varphi}_j{}^{(i+1)}$ for $i \neq n$ or $\boldsymbol{\varphi}_j{}^{(n)} = 0$ for $i = n$. Since j and i are arbitrarily selected, (2.7) holds for any natural frequency with respect to any inertance b_i, which means that the natural frequencies of the MDOF system can always be reduced by increasing the inertance of any inerter.

Note that for a discrete vibration system, $\lambda_j > 0$, $j = 1, \ldots, n$ always holds (if $\lambda_j = 0$, the vibration system reduces to a lower degree-of-freedom system), then the necessary and sufficient condition for $\frac{\partial \lambda_j}{\partial b_i} \leq 0$ is

$$\frac{\partial \mathbf{M}}{\partial b_i} \geq 0. \tag{2.14}$$

Thus, one obtains the following proposition:

Proposition 4 *1. The natural frequencies of the MDOF system shown in Fig. 2.5 can always be reduced by increasing the inertance of any inerter.*

2. The natural frequencies of any MDOF system can be reduced by an inerter if the inertial matrix satisfies (2.14).

Remark 2.2 The second conclusion in Proposition 4 means that the vibration systems of which the natural frequencies can be reduced by using an inerter are not restricted to the "uni-axial" MDOF system shown in Fig. 2.5, but any MDOF system satisfying (2.14), such as full-car suspension systems (Smith and Wang 2004), train suspension systems (Wang and Liao 2009; Wang et al. 2011; Jiang et al. 2012), buildings (Wang et al. 2010), etc.

Remark 2.3 Proposition 4 is easy to interpret physically. For a small increment of inertance ε_{b_i} of a particular inerter b_i, one obtains

$$\mathbf{M} = \mathbf{M}_0 + \varepsilon_{b_i} \frac{\partial \mathbf{M}}{\partial b_i}, \tag{2.15}$$

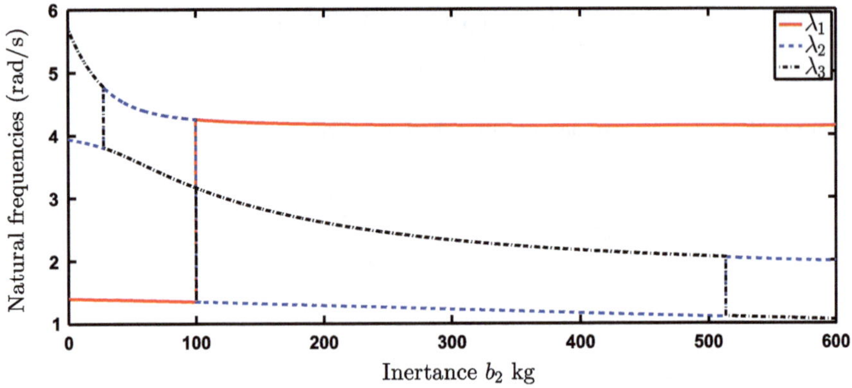

Fig. 2.6 The permutation of natural frequencies of a three-degree-of-freedom system with $m_i = 100$ kg, $k_i = 1000$ N/m, $i = 1, 2, 3$ and $b_1 = b_3 = 0$ kg, $b_2 \in [0, 600]$ kg

where $\mathbf{M_0}$ is the original inertial matrix. Sine $\frac{\partial \mathbf{M}}{\partial b_i}$ is positive semidefine, (2.15) can be interpreted as increasing the mass of the whole system, which will surely result in the reduction of natural frequencies.

Note that from Proposition 4, it seems that any natural frequency of an MDOF system will be reduced if an inerter with a relatively large value of inertance is inserted since the added inertance can always be viewed as an integration of small increments. However, this is not always true since there exist permutations of two particular natural frequencies if the divergence between two eigenvalues of the original system is not large enough or the increment of inertance ε_{b_i} is not small enough. Figure 2.6 shows the permutation of the natural frequencies of a three-degree-of-freedom system. As shown in Fig. 2.6, if one denotes the eigenvalues in the order of $\lambda_1 \geq \lambda_2 \geq \ldots \geq \lambda_n$ all the time, the $\lambda_i, i = 1, \ldots, n$, will always decrease when the inertance increases. Hence, in the following sections, the eigenvalues are always sorted in a descending order unless otherwise stated.

Remark 2.4 Note that the equality sign can be achieved for some natural frequencies of a particular system. This means that for some particular system, it is possible to reduce part of natural frequencies while maintaining others unchanged. This fact can be demonstrated by using a Two DOF system as shown in Fig. 2.7. If $m_1 = m_2 = m$, $k_1 = k_3 = k$, $b_1 = b_3 = b$, then the natural frequencies of the system are

$$\omega_{n1} = \sqrt{\frac{k}{m+b}}, \tag{2.16}$$

$$\omega_{n2} = \sqrt{\frac{k+2k_2}{m+b+2b_2}}. \tag{2.17}$$

It is clear that increasing b_2 can reduce ω_{n2} but cannot reduce ω_{n1}.

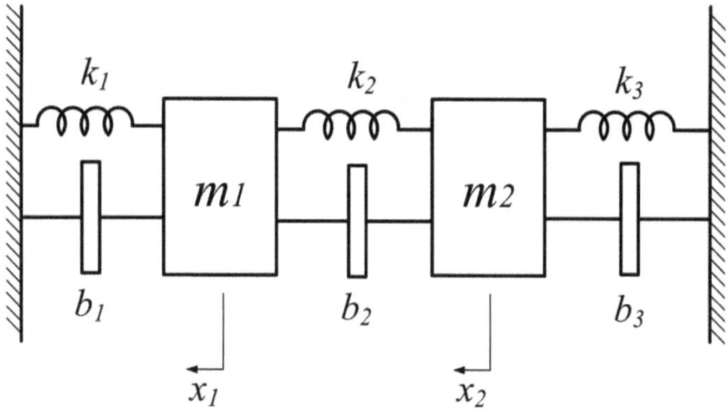

Fig. 2.7 A special TDOF system

2.6 Influence of the Inerter Position on the Natural Frequencies

The fact that inerter can reduce the natural frequencies of any MDOF system satisfying (2.14) has been demonstrated. However, for an MDOF system such as the "uni-axial" MDOF system shown in Fig. 2.5, the influence of inerter position on a specific natural frequency is still unknown. In particular, a practical problem is: for a specific natural frequency such as the largest natural frequency, where is the most efficient position to insert an inerter so that the largest reduction will be achieved? A TDOF system shown in Fig. 2.4 will be investigated in detail and analytical solutions will be derived for the TDOF system.

Considering (2.13) with $n = 2$, one obtains

$$\frac{\partial \lambda_j}{\partial b_1} = -\lambda_j \left(\varphi_j^{(1)} - \varphi_j^{(2)}\right)^2, \tag{2.18}$$

$$\frac{\partial \lambda_j}{\partial b_2} = -\lambda_j \left(\varphi_j^{(2)}\right)^2, \tag{2.19}$$

where $j = 1, 2$.

For a small increment of inertance, to compare the efficiency of reducing natural frequencies in terms of b_1 and b_2, it is equivalent to compare the absolute values of the derivatives in (2.18) and (2.19). Then, the following proposition can be derived.

Proposition 5 *For a small increment of inertance and for a specific λ_j, $j = 1, 2$, it is more efficient to increase b_1 than b_2 if*

$$\frac{k_1}{2m_1 + b_1} < \lambda_{j0} < \frac{k_1}{b_1}, \tag{2.20}$$

or

$$\lambda_{j0} > \frac{k_2}{m_2 + b_2}, \; or \; \lambda_{j0} < \frac{k_2}{m_2 + b_2 + 2m_1}. \tag{2.21}$$

It is more efficient to increase b_2 than b_1 if

$$\lambda_{j0} > \frac{k_1}{b_1}, \; or \; \lambda_{j0} < \frac{k_1}{b_1 + 2m_1}, \tag{2.22}$$

or

$$\frac{k_2}{m_2 + b_2 + 2m_1} < \lambda_{j0} < \frac{k_2}{m_2 + b_2}, \tag{2.23}$$

where λ_{j0}, $j = 1, 2$ denote the eigenvalues of the original system.

Proof Considering (2.6), one obtains

$$\varphi_j{}^{(1)} - \varphi_j{}^{(2)} = \frac{\lambda_j m_1}{k_1 - \lambda_j(m_1 + b_1)} \varphi_j{}^{(2)}, \tag{2.24}$$

$$= \frac{k_2 - \lambda_j(m_1 + m_2 + b_2)}{\lambda_j m_1} \varphi_j{}^{(2)}, \tag{2.25}$$

where $j = 1, 2$, and (2.24) is obtained by checking the first row of (2.6) and (2.25) is obtained by summing the first and second rows of (2.6).

Note that

$$\left|\frac{\partial \lambda_j}{\partial b_1}\right| - \left|\frac{\partial \lambda_j}{\partial b_2}\right| = \lambda_j \left((\varphi_j{}^{(1)} - \varphi_j{}^{(2)})^2 - (\varphi_j{}^{(2)})^2 \right).$$

Substituting (2.24) and (2.25), separately, one obtains the conditions in Proposition 5. □

Note that (2.20) and (2.21), (2.22) and (2.23) are equivalent, because (2.24) and (2.25) are equivalent. Proposition 5 is only applied to the case that the increment of inertance is small, as it is obtained by comparing the slopes of the tangent lines as shown in the proof of Proposition 5. If large increments of inertance are allowed for a given system that can be modeled as Fig. 2.4 and no inerter is employed in the original system, the question that which is more efficient in terms of b_1 and b_2 will be investigated as follows.

To answer this question, one needs to check two situations, where $b_2 = 0$ or $b_1 = 0$, respectively. If $b_2 = 0$, $b_1 = b$, from (2.4) and (2.5), one has

$$\omega_{n1} = \sqrt{\frac{(m_1 + m_2)k_1 + m_1 k_2 + k_2 b_1 - \sqrt{((m_1 + m_2)k_1 - m_1 k_2 - b_1 k_2)^2 + 4k_1 k_2 m_1^2}}{2(m_1 m_2 + (m_1 + m_2)b_1)}},$$

$$\omega_{n2} = \sqrt{\frac{(m_1 + m_2)k_1 + m_1 k_2 + k_2 b_1 + \sqrt{((m_1 + m_2)k_1 - m_1 k_2 - b_1 k_2)^2 + 4k_1 k_2 m_1^2}}{2(m_1 m_2 + (m_1 + m_2)b_1)}}.$$

If $b_1 = 0$, $b_2 = b$, one has

$$\omega'_{n1} = \sqrt{\frac{(m_1+m_2)k_1 + m_1k_2 + k_1b_2 - \sqrt{((m_1+m_2)k_1 - m_1k_2 + b_2k_1)^2 + 4k_1k_2m_1^2}}{2(m_1m_2+m_1b_2)}},$$

$$\omega'_{n2} = \sqrt{\frac{(m_1+m_2)k_1 + m_1k_2 + k_1b_2 + \sqrt{((m_1+m_2)k_1 - m_1k_2 + b_2k_1)^2 + 4k_1k_2m_1^2}}{2(m_1m_2+m_1b_2)}}.$$

The above question can be answered by comparing ω_{n1} and ω_{n2} with ω'_{n1} and ω'_{n2}, respectively. Thus, one has the following proposition.

Proposition 6 *Denote*

$$b_0 = \frac{k_1m_2(2m_1k_2 - (2m_1+m_2)k_1)}{(k_2-k_1)(m_1k_2-(m_1+m_2)k_1)}.$$

For the larger natural frequency ω_{n2}:
If $k_2 \le (1+\frac{m_2}{m_1})k_1$, b_1 is more efficient than b_2;
If $k_2 > (1+\frac{m_2}{m_1})k_1$, b_1 is more efficient in $[0, b_0]$; b_2 is more efficient in $[b_0, +\infty)$.
For the smaller natural frequency ω_{n1}:
If $k_2 > (1+\frac{m_2}{2m_1})k_1$, b_1 is more efficient than b_2;
If $k_1 \le k_2 \le (1+\frac{m_2}{2m_1})k_1$, b_2 is more efficient in $[0, b_0]$; b_1 is more efficient in $[b_0, +\infty)$;
If $k_2 < k_1$, b_2 is more efficient than b_1.

Proof Denote $b_1 = b_2 = b$,

$$\begin{aligned} d_1 &= 2(m_1m_2+m_1b), \\ d_2 &= 2(m_1m_2+(m_1+m_2)b), \\ d_3 &= (m_1+m_2)k_1 + m_1k_2 + k_2b, \\ d_4 &= (m_1+m_2)k_1 + m_1k_2 + k_1b, \\ d_5 &= \sqrt{(bk_2 + m_1k_2 - (m_1+m_2)k_1)^2 + 4k_1k_2m_1^2}, \\ d_6 &= \sqrt{(bk_1 - m_1k_2 + (m_1+m_2)k_1)^2 + 4k_1k_2m_1^2}, \end{aligned}$$

and

$$F_1(b) = \omega_{n1}^2 - {\omega'}_{n1}^2 = \frac{d_1d_3 - d_2d_4 - d_1d_5 + d_2d_6}{d_1d_2},$$

$$F_2(b) = \omega_{n2}^2 - {\omega'}_{n2}^2 = \frac{d_1d_3 - d_2d_4 + d_1d_5 - d_2d_6}{d_1d_2}.$$

Also denote

$$b_0 = \frac{k_1 m_2 (2m_1 k_2 - (2m_1 + m_2)k_1)}{(k_2 - k_1)(m_1 k_2 - (m_1 + m_2)k_1)}.$$

By direct calculation, it can be easily verified that both $F_1(b) = 0$ and $F_2(b) = 0$ have solutions at 0 and b_0. However, note that $F_1(b)$ and $F_2(b)$ cannot be zero at the same time if $b \neq 0$, thus $F_1(b_0) = 0$ and $F_2(b_0) = 0$ cannot hold simultaneously. Particularly, since $b > 0$, one is more interested in the cases that $k_2 \in [k_1, (1 + m_2/(2m_1))k_1]$ and $k_2 \in [(1 + m_2/m_2)k_1, \infty)$, where $b_0 \geq 0$.

Next, it is shown that the positive value of b_0 in $k_2 \in [(1 + m_2/m_2)k_1, \infty)$ belongs to $F_2(b) = 0$ and the other one belongs to $F_1(b) = 0$. Denote

$$\Delta_2 = m_1 k_2 - (m_1 + m_2)k_1,$$
$$\Delta_1^2 = \Delta_2^2 + 4k_1 k_2 m_1^2.$$

Then

$$d_5 = \sqrt{bk_2^2 + 2\Delta_2 k_2 b + \Delta_1^2} = k_2 b + \Delta_2 + \frac{2k_1 m_1^2}{b} + O\left(\frac{1}{b^2}\right),$$
$$d_6 = \sqrt{bk_1^2 - 2\Delta_2 k_1 b + \Delta_1^2} = k_1 b - \Delta_2 + \frac{2k_2 m_1^2}{b} + O\left(\frac{1}{b^2}\right).$$

Hence, one has

$$\begin{aligned} F_2(b) &= \frac{d_1 d_3 - d_2 d_4 + d_1 d_5 - d_2 d_6}{d_1 d_2} \\ &= \frac{\Delta_2 (4b^2 + 4(m_1 + m_2)b + 4m_1 m_2)}{d_1 d_2} - \\ &\quad \frac{4m_1 (m_2 k_1 - m_1 (k_1 m_1 - k_2 (m_1 + m_2)))}{d_1 d_2} + O\left(\frac{1}{b}\right). \end{aligned}$$

Note that if $\Delta_2 < 0$ and $k_2 > k_1$, or $k_1 < k_2 < (1 + m_2/m_1)k_1$, $F_2(b)$ is always negative by omitting the higher order item $O\left(\frac{1}{b}\right)$. This indicates that if $k_2 < (1 + m_2/m_1)k_1$, then $F_2(b) = 0$ only has the trivial solution 0, while if $k_2 \geq (1 + m_2/m_1)k_1$, then $F_2(b) = 0$ has solutions at 0 and b_0. Consequently, if $k_2 < (1 + m_2/m_1)k_1$, then $F_1(b) = 0$ has roots at 0 and b_0, while if $k_2 \geq (1 + m_2/m_1)k_1$, then $F_1(b) = 0$ only has a trivial solution 0.

Besides, since

$$\begin{aligned} F_1(b) &= \frac{d_1 d_3 - d_2 d_4 - d_1 d_5 + d_2 d_6}{d_1 d_2}, \\ &= \frac{4m_1 (m_1 + m_2)(k_1 - k_2)b - 4m_1 (m_1^2 (k_1 - k_2) - m_2 k_1 (m_1 + m_2)) - O\left(\frac{1}{b}\right)}{d_1 d_2}, \end{aligned}$$

by the relationship of the coefficients and the roots of $F_1(b)$ and $F_2(b)$, one has
If $k_2 > (1 + m_2/m_1)k_1$, $F_1(b) \leq 0$ and $F_2(b) \leq 0$ for $b \in [0, b_0]$, $F_2(b) > 0$ for $b \in (b_0, \infty)$;
If $(1 + m_2/(2m_1))k_1 \leq k_2 \leq (1 + m_2/m_1)k_1$, $F_1(b) < 0$ and $F_2(b) < 0$;
If $k_1 \leq k_1 < (1 + m_2/(2m_1))k_1$, $F_1(b) \geq 0$ for $b \in [0, b_0]$, $F_1(b) < 0$ for $b \in (b_0, \infty)$, and $F_2(b) < 0$;
If $k_2 < k_1$, $F_1(b) > 0$ and $F_2(b) < 0$.

Thus, Proposition 6 and the four cases shown in Fig. 2.8 have been proved. □

Proposition 6 has addressed four cases, which are $k_2 > (1 + m_2/m_1)k_1$, $(1 + m_2/(2m_1))k_1 \leq k_2 \leq (1 + m_2/m_1)k_1$, $k_1 \leq k_2 < (1 + m_2/(2m_1))k_1$, $k_2 \leq k_1$. A numerical example is performed with $m_1 = m_2 = 100$ kg, $k_1 = 1000$ N/m and k_2 chosen as 2500, 1800, 1300, 500 N/m corresponding to the four cases in Proposition 6. The results are shown in Fig. 2.8, where one sees that in terms of the larger natural frequency, although for small increment of inertance (about 0–250 kg) b_1 is more efficient than b_2, for large increment of inertance, b_2 tends to be more efficient than b_1.

Note that the above discussion is based on TDOF systems. For a general MDOF system, a similar argument as in Proposition 5 can be employed to determine the efficiency of the position of inerter by comparing the absolute values of the derivatives. For example, consider a six-degree-of-freedom system with $m_i = 100$ kg, $i = 1, \ldots, 6$, and $k_1 = 1000$ N/m, $k_2 = 1000$ N/m, $k_3 = 2000$ N/m, $k_4 = 2000$ N/m, $k_5 = 3000$ N/m, $k_6 = 3000$ N/m. The objective is to find out the most efficient position to insert an inerter so that largest reduction of the largest natural frequency will be achieved. By direct calculation, one obtains $\left|\frac{\partial \lambda_1}{\partial b_i}\right|$, $i = 1, \ldots, 6$ as 2.759×10^{-4}, 0.0134, 0.1559, 0.8571, 1.5999, 0.4043, respectively. Note that $\left|\frac{\partial \lambda_1}{\partial b_5}\right|$ possesses the largest value. Hence, the position between m_5 and m_6 would be the most efficient position to insert an inerter, which is consistent with the simulation shown in Fig. 2.9. Another method to find the most efficient position is by using Gershgorin's Theorem (Horn and Johnson 1988), which shows that the largest absolute row sums is an upper bound of the largest eigenvalue. Hence, an efficient way to reduce the largest natural frequency is to insert the inerter between the mass m_j and m_{j+1} or m_{j-1} and m_j, where the jth absolute row sum of $\mathbf{M}^{-1}\mathbf{K}$ is the largest absolute row sum of $\mathbf{M}^{-1}\mathbf{K}$. Taking the same six-degree-of-freedom system as an example, one obtains

$$\mathbf{M}^{-1}\mathbf{K} = \begin{bmatrix} 10 & -10 & 0 & 0 & 0 & 0 \\ -10 & 20 & -10 & 0 & 0 & 0 \\ 0 & -10 & 30 & -20 & 0 & 0 \\ 0 & 0 & -20 & 40 & -20 & 0 \\ 0 & 0 & 0 & -20 & 50 & -30 \\ 0 & 0 & 0 & 0 & -30 & 60 \end{bmatrix}.$$

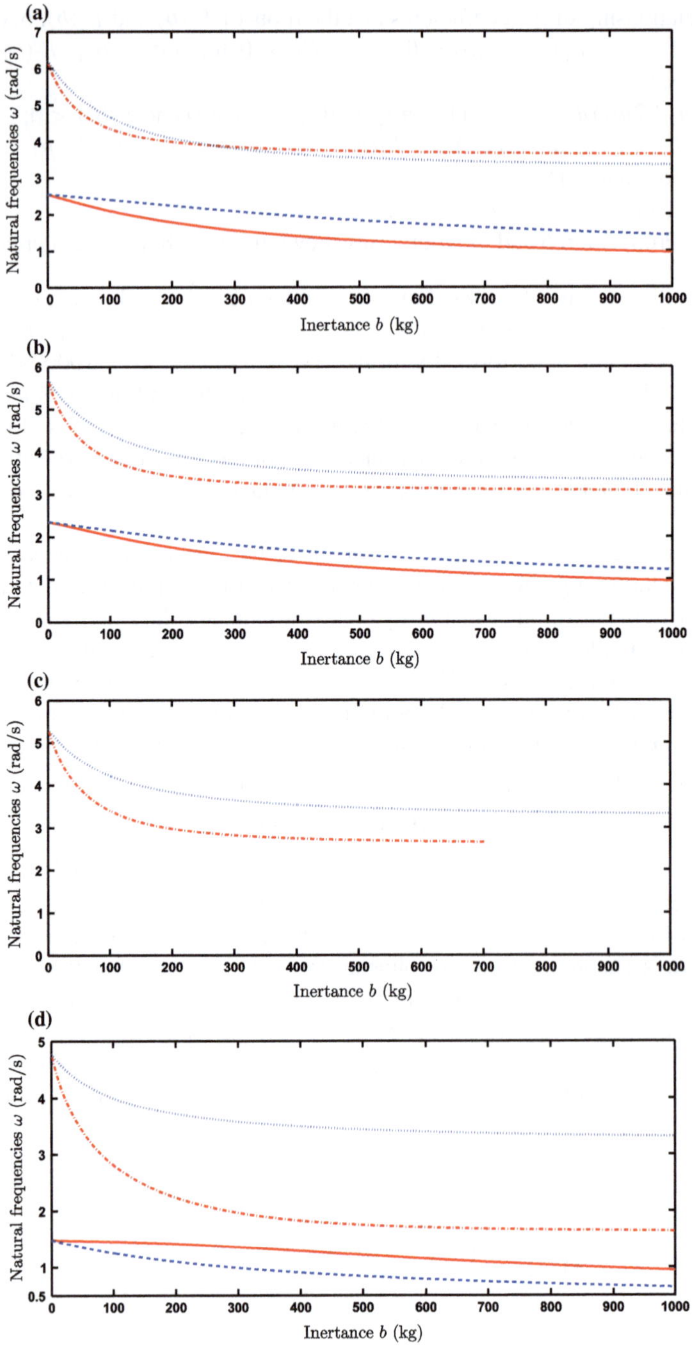

Fig. 2.8 The natural frequencies of the TDOF system. **a** $k_2 > (1 + m_2/m_1)k_1$; **b** $(1 + m_2/(2m_1))k_1 \leq k_2 \leq (1 + m_2/m_1)k_1$; **c** $k_1 \leq k_2 < (1 + m_2/(2m_1))k_1$; **d** $k_2 \leq k_1$. The red solid line: ω_{n1}; the blue dashed line: ω'_{n1}; the red dash-dot line: ω_{n2}; the blue dotted line: ω'_{n2}

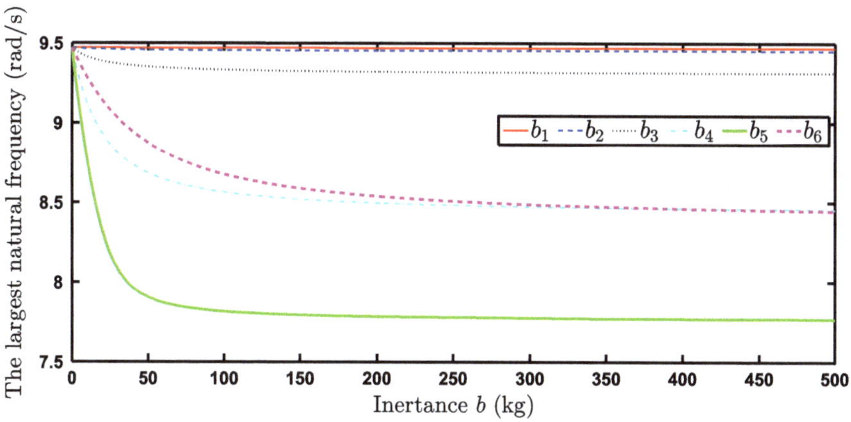

Fig. 2.9 The largest natural frequency of a six-degree-of-freedom system

The absolute row sums of $\mathbf{M}^{-1}\mathbf{K}$ are 20, 40, 60, 80, 100, and 90. Thus, one concludes that the optimal way is to insert an inerter between m_5 and m_6, which is consistent with the simulation shown in Fig. 2.9 as well.

2.7 Design Procedure and Numerical Example

The problem of reducing the largest natural frequency of a vibration system is considered in this section, where the efficiency of inerter in reducing natural frequencies will be quantitatively shown.

For the largest natural frequency, considering (2.8) and (2.9), one obtains

$$\frac{\partial \Phi_{ij}}{\partial b_i} \leq 0, \text{ and } \frac{\partial^2 \lambda_j}{\partial b_i^2} \geq 0.$$

Table 2.1 Structure model parameters

Floor masses (kg)	Stiffness coefficients (kN/m)
$m_1 = 5897$	$k_1 = 19059$
$m_2 = 5897$	$k_2 = 24954$
$m_3 = 5897$	$k_3 = 28621$
$m_4 = 5897$	$k_4 = 29093$
$m_5 = 5897$	$k_5 = 33732$
$m_6 = 6800$	$k_6 = 232$

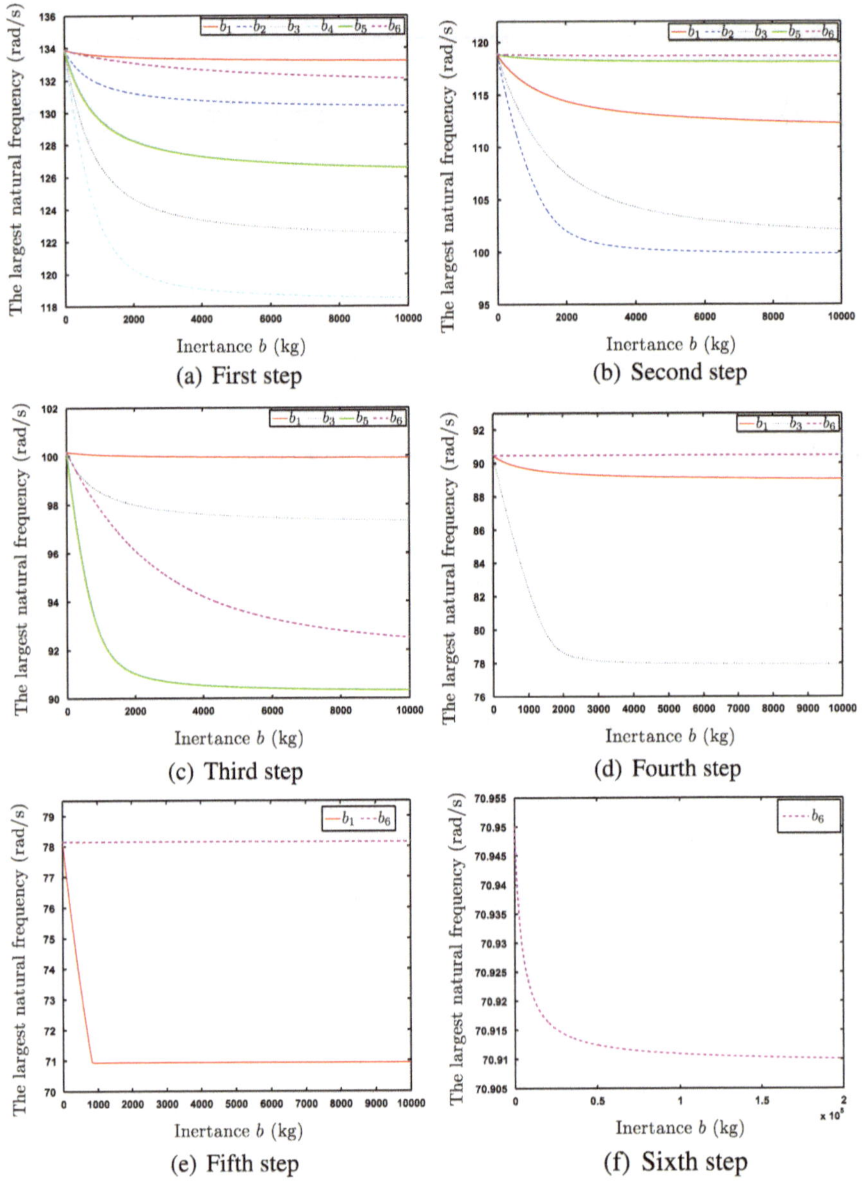

Fig. 2.10 Procedures. **a** first step; **b** second step: $b_4 = 5000$ kg; **c** third step: $b_4 = 5000$ kg, $b_2 = 5000$ kg; **d** fourth step: $b_4 = 5000$ kg, $b_2 = 5000$ kg, $b_5 = 5000$ kg; **e** fifth step: $b_4 = 5000$ kg, $b_2 = 5000$ kg, $b_5 = 5000$ kg, $b_3 = 3000$; **f** sixth step: $b_4 = 5000$ kg, $b_2 = 5000$ kg, $b_5 = 5000$ kg, $b_3 = 3000$, $b_1 = 1000$ kg

Table 2.2 Procedures and results

Steps	Inertance (kg)	$\omega_{\max}$ (rad/s)	Percentages (%)
1st	$b_4 = 5000$	118.89	(11.22)
2th	$b_4 = 5000$ $b_2 = 5000$	100.19	(25.18)
3th	$b_4 = 5000$ $b_2 = 5000$ $b_5 = 5000$	90.49	(32.43)
4th	$b_4 = 5000$ $b_2 = 5000$ $b_5 = 5000$ $b_3 = 3000$	78.15	(41.64)
5th	$b_4 = 5000$ $b_2 = 5000$ $b_5 = 5000$ $b_3 = 3000$ $b_1 = 1000$	70.95	(47.02)
6th	$b_4 = 5000$ $b_2 = 5000$ $b_5 = 5000$ $b_3 = 3000$ $b_1 = 1000$ $b_6 = 1 \times 10^5$	70.91	(47.05)

Note that $\Phi_{ij} \geq 0$ and the equality is achieved with $\varphi_j^{(i)} = \varphi_j^{(i+1)}$ when $i \neq n$, or $\varphi_j^{(n)} = 0$ when $i = n$, which means that for a specific inerter b_i, $i = 1, \ldots, n$, the largest natural frequency will always be reduced by increasing the inertance until the two masses connected by inerter b_i are rigidly connected.

In what follows, an intuitive and simple approach to lowering the largest natural frequency for a given structure is illustrated by inserting the inerters one by one, where the inerter in each step is placed at the most efficient position. Here, a procedure is presented to reduce the largest natural frequency of a structure discussed in Kelly et al. (1987), Ramallo et al. (2002) with parameters given in Table 2.1. Note that the largest natural frequency $\omega_{\max}$ of this structure is 133.91 rad/s. The procedure to reduce $\omega_{\max}$ is shown in Fig. 2.10 and Table 2.2.

Procedure description:

Step 1 Figure 2.10a shows that b_4 is the most efficient regarding the original system and for $b_4 > 5000$ kg, $\omega_{\max}$ decreases slightly, and hence $b_4 = 5000$ kg is selected;

Step 2 Figure 2.10b shows that b_2 is the most efficient regarding the original system and b_4 and $b_2 > 5000$ kg, $\omega_{\max}$ decreases slightly, and hence $b_2 = 5000$ kg is selected;

Step 3–Step 6 Similarly, from Fig. 2.10c to f, $b_5 = 5000$ kg, $b_3 = 3000$ kg, $b_1 = 1000$ kg, and $b_6 = 1 \times 10^5$ kg are selected, respectively.

Note that the above-illustrated approach is not optimal as the natural frequencies of a system can always be reduced by enlarging the inertance until the inertial matrix **M** became singular, where all the natural frequencies become zero. However, the efficiency of inerter in reducing natural frequencies can be clearly demonstrated by this approach. As shown in Table 2.2, attenuation about 47.05% has been obtained. It is worth pointing out that the required inertance for b_6 is 1×10^5 kg, which is quite large. However, the reduction of largest natural frequency is only improved by 0.03%. If the cost factor is considered in practice, b_6 can be omitted. In this way, only five inerters are employed.

2.8 Conclusions

This chapter has investigated the influence of inerter on the natural frequencies of vibration systems. By algebraically deriving the natural frequencies of an SDOF system and a TDOF system, the fact that inerter can reduce the natural frequencies of these systems has been clearly demonstrated. To reveal the influence of inerter on the natural frequencies of a general system, an MDOF system has been considered. Sensitivity analysis has been performed on the natural frequencies and mode shapes to demonstrate that any increment of the inertance of any inerter in an MDOF system results in the reduction of the natural frequencies. To that end, the effectiveness of inerter in reducing natural frequencies of a general vibration system has been clearly demonstrated. Finally, the influence of the inerter position has been investigated and a simple design procedure has been proposed to verify the efficiency of inerter in reducing the largest natural frequencies of vibration systems. The simulation result has shown that more than 47% reduction can be obtained with only five inerters employed in a six-degree-of-freedom vibration system.

References

Beeby, S. P., Tudor, M. J., & White, N. M. (2006). Energy harvesting vibration sources for microsystems applications. *Measurement Science and Technology*, *17*(12), R175–R195.

Chen, M. Z. Q., Hu, Y., & Du, B. (2012). Suspension performance with one damper and one inerter. In *The 24th Chinese Control and Decision Conference, Taiyuan, China* (pp. 3551–3556).

Horn, R. A., & Johnson, C. R. (1988). *Matrix analysis*. New York: Cambridge University Press.

Jiang, J. Z., Matamoros-Sanchez, A. Z., Goodall, R. M., & Smith, M. C. (2012). Passive suspensions incorporating inerters for railway vehicles. *Vehicle System Dynamics*, *50*(sup1), 263–276.

Kelly, J. M., Leitmann, G., & Soldatos, A. G. (1987). Robust control of base-isolated structures under earthquake excitation. *Journal of Optimization Theory and Applications*, *53*, 159–180.

Lee, I. W., & Kim, D. O. (1999). Natural frequency and mode shape sensitivities of damped systems: Part I, distinct natural frequencies. *Journal of Sound and Vibration*, *223*(3), 399–412.

Lin, J., & Parker, R. G. (1999). Sensitivity of planetary gear natural frequencies and vibration modes to model parameters. *Journal of Sound and Vibration*, *228*(1), 109–128.

Ramallo, J. C., Johnson, E. A., & Spencer, B. F. (2002). Smart base isolation systems. *Journal of Engineering Mechanics (ASCE)*, *128*(10), 1088–1099.

Smith, M. C. (2002). Synthesis of mechanical networks: The inerter. *IEEE Transactions on Automatic Control*, *47*(1), 1648–1662.

Smith, M. C., & Wang, F. C. (2004). Performance benefits in passive vehicle suspensions employing inerters. *Vehicle System Dynamics*, *42*(4), 235–257.

Thomson, W. T. (1993). *Theory of vibration with applications* (4th ed.). Englewood Cliffs, NJ: Prentice-Hall.

Tse, F. S., Morse, I. E., & Hinkle, R. T. (1979). *Mechanical vibrations*. Maruzen Asian Edition: Allyn and Bacon.

Wang, F. C., & Liao, M. K. (2009). The lateral stability of train suspension systems employing inerters. *Vehicle System Dynamics*, *48*(5), 619–643.

Wang, F. C., Hong, M. F., & Chen, C. W. (2010). Building suspensions with inerters. *Proceedings of the Institution of Mechanical Engineers, Part C: Journal of Mechanical Engineering Science*, *224*(8), 1605–1616.

Wang, F. C., Hsieh, M. R., & Chen, H. J. (2011). Stability and performance analysis of a full-train system with inerters. *Vehicle System Dynamics*, *50*(4), 545–571.

Yu, Y., Naganathan, N. G., & Dukkipati, R. V. (2001). A literature review of atuomotive vehilce engine mounting systems. *Mechanism and Machine Theory*, *36*(1), 123–142.

Zhao, J., & DeWolf, J. T. (1999). Sensitivity study for vibrational parameters used in damage detection. *Journal of Structural Engineering*, *125*(4), 410–416.

Chapter 3
Inerter-Based Isolation System

Abstract This chapter is concerned with the problem of analysis and optimization of the inerter-based isolators based on a "uni-axial" single-degree-of-freedom isolation system. In the first part, in order to gain an in-depth understanding of inerter from the prospective of vibration, the frequency responses of both parallel-connected and series-connected inerters are analyzed. In the second part, three other inerter-based isolators are introduced and the tuning procedures in both the H_∞ optimization and the H_2 optimization are proposed in an analytical manner. The achieved H_2 and H_∞ performance of the inerter-based isolators is superior to that achieved by the traditional dynamic vibration absorber (DVA) when the same inertance-to-mass (or mass) ratio is considered. Moreover, the inerter-based isolators have two unique properties, which are more attractive than the traditional DVA: first, the inertance-to-mass ratio of the inerter-based isolators can easily be larger than the mass ratio of the traditional DVA without increasing the physical mass of the whole system; second, there is no need to mount an additional mass on the object to be isolated.

Keywords Vibration isolation · H_∞ optimization · H_2 optimization · Transmissibility · Analytical analysis.

3.1 Introduction

In this chapter, to further investigate the influence of inerter on vibration systems, the performance of the inerter-based isolators based on a "uni-axial" single-degree-of-freedom isolation system is studied. First, to gain an in-depth understanding of inerter from the perspective of vibration, the frequency responses of both parallel-connected and series-connected inerters are analyzed. It is shown that an extra invariant point, which is independent of the damping ratio, can be introduced by using the series-connected inerter. Then, to further tune the invariant points, three other inerter-based isolators, each of which incorporates a spring, a damper, and an inerter, are proposed. To facilitate the practical application, the optimal parameters of the inerter-based isolators in both H_∞ optimization and H_2 optimization are analytically derived. An analytical method is employed to calculate the H_2 norm performance measures of

M. Z. Q. Chen and Y. Hu, *Inerter and Its Application in Vibration Control Systems*, https://doi.org/10.1007/978-981-10-7089-1_3

the inerter-based isolators in this paper. In addition, the comparisons of the H_2 and H_∞ performances between the inerter-based isolators and the traditional DVA show the superiority of the inerter-based isolators. Two properties make the inerter-based isolators potentially more attractive than the traditional DVA: first, a relatively large inertance can easily be obtained without increasing the physical mass of the whole system (Smith 2002); second, there is no need to mount an additional mass on the object to be isolated, as an inerter is a built-in component in the inerter-based isolators.

3.2 Preliminary

Vibration isolation is one of the most common vibration control categories, where two situations are commonly encountered in terms of the vibrating source. The first one is to protect the object from vibrating environment. For example, equipments may be mounted on an isolator to be protected from an environment characterized by severe shock or vibration. The other situation is the isolation of the vibrating source. For example, a machine creating significant vibration during operation may be supported upon isolators such that other parts of the systems or other machines are less influenced (Piersol and Paez 2010).

In Fig. 3.1, a "uni-axial" isolation system is shown, where the mass m is the object to be isolated, the mass m_f is the foundation, and $Q(s)$ is the isolator to be designed. The first situation discussed above can be described as the displacement transmissibility problem and the later one is the force transmissibility problem (Carrella et al. 2012). In some cases, both tasks have to be addressed simultaneously (Rivin 2003).

For the displacement transmissibility problem with harmonic inputs, the absolute transmissibility from x_2 to x_1 denoted as μ_x can be obtained as

$$\mu_x = \frac{\mid x_1 \mid}{\mid x_2 \mid} = \frac{\mid Q(j\omega)j\omega \mid}{\mid Q(j\omega)j\omega - m\omega^2 \mid}, \tag{3.1}$$

where ω is the input frequency and $F = 0$.

For the force transmissibility problem with harmonic inputs, denoting F_f as the force transformed from the object to the foundation, the absolute transmissibility from F to F_f denoted as μ_F can be obtained as

$$\mu_F = \frac{\mid F_f \mid}{\mid F_0 \mid} = \frac{m_f}{m + m_f} \frac{\mid Q(j\omega)j\omega \mid}{\mid Q(j\omega)j\omega - m_{ef}\omega^2 \mid}, \tag{3.2}$$

where $F_i = F_f$.

Comparing (3.1) and (3.2), one concludes that $\mu_F = \mu_x$ if only if $m_f = \infty$, which means that for any passive isolator, the force transmissibility problem is equivalent to the displacement transmissibility problem if only if the mass of the foundation is sufficiently larger than that of the object.

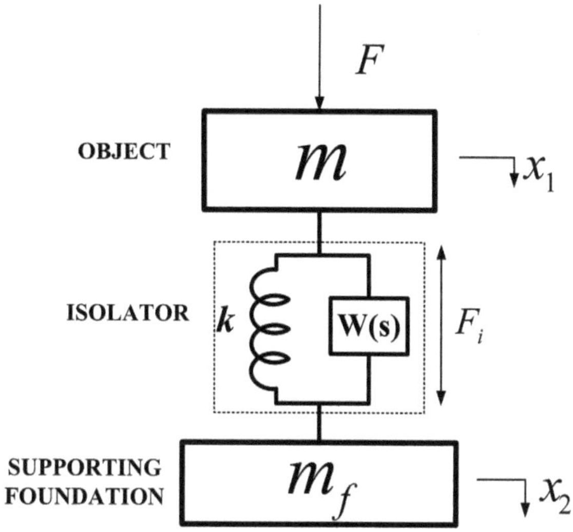

Fig. 3.1 Uni-axial vibration isolation system

For brevity, in this chapter, the assumption that $m_f = \infty$ is made and the *absolute displacement transmissibility* and the *absolute force transmissibility* are identically treated as

$$\mu = \frac{|F_i|}{|F|} = \frac{|x_1|}{|x_2|} = \frac{|Q(\mathrm{j}\omega)\mathrm{j}\omega|}{|Q(\mathrm{j}\omega)\mathrm{j}\omega - m\omega^2|}, \tag{3.3}$$

where F is the force imposed on the object m, F_i is the force generated by the isolator, x_1 and x_2 are the displacements of the object and the foundation, respectively. $Q(\mathrm{j}\omega)$ is obtained by replacing the Laplace variable s in $Q(s)$ with $\mathrm{j}\omega$, where j is a complex variable with $\mathrm{j}^2 = -1$ and $Q(s)$ is the admittance of the isolator, i.e., the ratio of the applied force F_i over the relative velocity $\dot{x}_1 - \dot{x}_2$ in Laplace domain.

As shown in Fig. 3.1, $Q(s) = \frac{k}{s} + W(s)$, where $W(s)$ denotes the admittances of passive networks consisting of finite inter-connections of springs, dampers, and inerters. For the traditional isolator without inerters, i.e., $W(s)$ is a damper, the transmissibility can be obtained as

$$\mu = \frac{|k + \mathrm{j}c\omega|}{|k - m\omega^2 + \mathrm{j}c\omega|} = \sqrt{\frac{(1-q^2)^2 + (2\zeta q)^2}{(1-q^2)^2 + (2\zeta q)^2}}, \tag{3.4}$$

where $q = \frac{\omega}{\omega_n}$, $\zeta = \frac{c}{c_r}$ with $c_r = 2\sqrt{mk}$.

The frequency response, i.e., transmissibility μ for the traditional isolator without inerters is shown in Fig. 3.2, where an invariant point independent of the damping ratio ζ is depicted. The H_∞ optimization aims to minimize the maximum magnitude of the frequency response, while the H_2 optimization aims to minimize the mean squared displacement of the object under random excitation (Cheung and Wong 2011b).

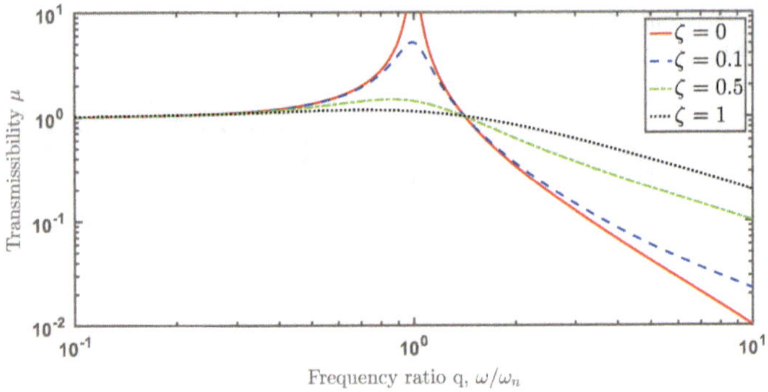

Fig. 3.2 Transmissibility μ for the traditional isolator without inerters

Table 3.1 $W(s)$ for configurations in Figs. 3.3 and 3.8, where s denotes the Laplace variable

$W_1(s) = bs + c$	$W_2(s) = \frac{1}{\frac{1}{c}+\frac{1}{bs}}$	$W_3(s) = \frac{1}{\frac{1}{\frac{k_1}{s}+c}+\frac{1}{bs}}$	$W_4(s) = \frac{1}{\frac{s}{k_1}+\frac{1}{bs}+\frac{1}{c}}$	$W_5(s) = \frac{1}{\frac{1}{bs+c}+\frac{s}{k_1}}$

In this chapter, five inerter-based isolators will be investigated, as shown in Figs. 3.3 and 3.8. Their admittances are summarized in Table 3.1.

To obtain a dimensionless representation, $\omega_n = \sqrt{\frac{k}{m}}$ and $c_r = 2\omega_n m = 2\sqrt{mk}$ are used to denote the natural frequency and the critical damping of the isolation system shown in Fig. 3.1 without $W(s)$, respectively. Also, $q = \frac{\omega}{\omega_n}$, $\zeta = \frac{c}{c_r}$, $\delta = \frac{b}{m}$, and $\lambda = \frac{k}{k_1}$ denote the frequency ratio, the damping ratio, the inertance-to-mass ratio, and the stiffness ratio, respectively.

For the considered configurations as shown in Figs. 3.3 and 3.8, the transmissibility μ can be obtained by substituting $Q_i(\mathrm{j}\omega) = \frac{k}{\mathrm{j}\omega} + W_i(\mathrm{j}\omega)$, $i = 1, \ldots, 5$, into (3.3), respectively, where $W_i(\mathrm{j}\omega)$ are given in Table 3.1 by replacing s with $\mathrm{j}\omega$.

3.3 Vibration Analysis for Two Simple Inerter-Based Isolators

This section is to analyze the fundamental properties of inerter from the perspective of vibration. Note that among all the applications of inerter, the main focus is to optimize some inerter-based mechanical networks possessing more complex structures than the conventional networks consisting of only springs and dampers. The proposed mechanical networks can be obtained either by using networks synthesis (Wang and Chan 2011; Wang et al. 2012, 2014) or by giving some fixed-structure networks (Smith and Wang 2004; Chen et al. 2014; Hu et al. 2014; Scheibe and Smith 2009; Marian and Giaralis 2014; Lazar et al. 2014; Dylejko and MacGillivray

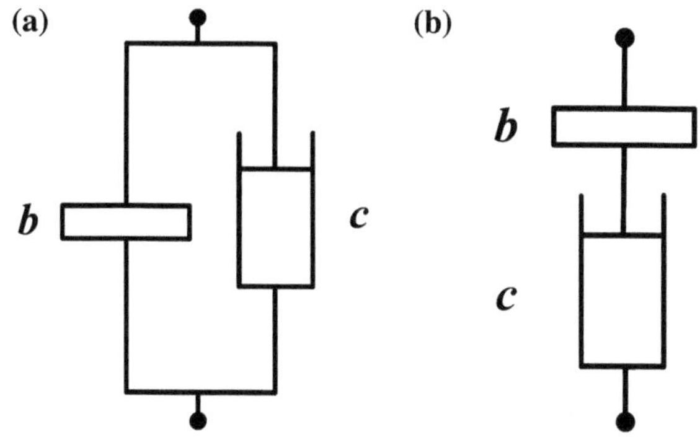

Fig. 3.3 Two simple configurations as $W(s)$ of the isolators in Fig. 3.1. **a** C1; **b** C2

2014). Although the benefits of using inerter can be effectively demonstrated by these complex inerter-based mechanical networks, some fundamental properties of inerter in vibration are overlooked due to the complexity of the structure. Consequently, it lacks in-depth understanding of inerter from the perspective of vibration. In Chen et al. (2014), the property that inerter can reduce vibration systems' natural frequencies is demonstrated. However, the influences of inerter on other aspects such as the invariant property in frequency domain are still unclear. This motivated the investigation of this section based on two simple inerter-based configurations, as shown in Fig. 3.3. The detailed analysis of the frequency responses of these configurations constitutes the main contribution of this section.

Analysis of C1

For this configuration, the transmissibility can be obtained as

$$\mu = \frac{| k - b\omega^2 + jc\omega |}{| k - (m+b)\omega^2 + jc\omega |} = \sqrt{\frac{(1-\delta q^2)^2 + (2\zeta q)^2}{(1-(1+\delta)q^2)^2 + (2\zeta q)^2}}. \tag{3.5}$$

Figure 3.4 shows the transmissibility μ with respect to different δ and ζ, where it is shown that an anti-resonant frequency (a particular frequency where minimum magnitude is obtained) and an invariant point (a particular frequency where the magnitude is independent of the damping ratio ζ) are introduced by using the parallel-connected inerter. For the undamped case, the anti-resonant frequency q_b can be obtained as $q_b = \sqrt{\frac{1}{\delta}}$, and the resonant frequency or natural frequency is $q_p = \sqrt{\frac{1}{1+\delta}}$. Note that the natural frequency q_p is a decreasing function with respect to δ, which is consistent with the result in Chen et al. (2014).

The transmissibility μ in (3.5) can be rewritten as

$$\mu = \sqrt{\frac{A\zeta^2 + B}{C\zeta^2 + D}},$$

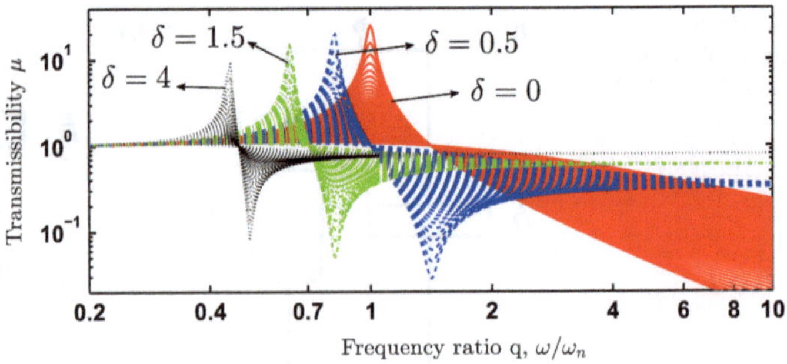

Fig. 3.4 Transmissibility μ for the configuration $C1$ when ζ ranges from 0.02 to 1.2

where $A = 4q^2$, $B = (1 - \delta q^2)^2$, $C = 4q^2$, and $D = (1 - (1 + \delta)q^2)^2$. To find the invariant points which are independent of damping, it requires

$$\frac{A}{C} = \frac{B}{D},$$

that is,

$$\frac{(1 - \delta q^2)^2}{(1 - (1 + \delta)q^2)^2} = 1.$$

Then, one obtains the nonzero invariant point q_i as

$$q_i = \sqrt{\frac{2}{1 + 2\delta}}.$$

Obviously, q_i is a decreasing function with respect to δ, which means that the parallel-connected inerter can effectively shift the invariant point left.

Figure 3.5 depicts the transmissibility μ of configuration $C1$ when $\delta = 1$ with some typical ζ. The magnitudes at the natural frequency q_p, the anti-resonant fre-

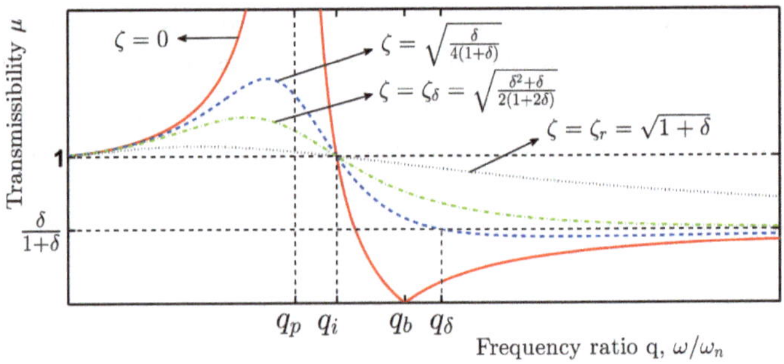

Fig. 3.5 Transmissibility μ for the configuration $C1$ when $\delta = 1$

quency q_b, and infinity can be obtained as

$$\mu|_{q=q_p} = \frac{1}{2}\sqrt{\frac{1}{\zeta^2(1+\delta)}+4}, \tag{3.6}$$

$$\mu|_{q=q_b} = 2\sqrt{\frac{1}{\frac{1}{\zeta^2\delta}+4}}, \tag{3.7}$$

$$\mu|_{q\to\infty} = \frac{\delta}{1+\delta}, \tag{3.8}$$

where $\mu|_{q=q_j}$ means the value of μ when $q = q_j$, j denotes p, b, or ∞.

From (3.6) and (3.7), it is clear that $\mu|_{q=q_p}$ is a decreasing function with respect to both δ and ζ, and $\mu|_{q=q_b}$ is an increasing function with respect to both δ and ζ, as shown in Fig. 3.4. From (3.7), one obtains that for the undamped case, i.e., $c = 0$ or $\zeta = 0$, $\mu|_{q=q_b} = 0$, the effect of "dynamic absorption" of vibration occurs, which is uncommon for single-degree-of-freedom systems (Rivin 2003).

Equation (3.8) shows that the transmissibility approaches to an asymptote at the level of $\frac{\delta}{1+\delta}$ when q tends to ∞. For a given δ, by solving the equation

$$\mu = \sqrt{\frac{(1-\delta q^2)^2+(2\zeta q)^2}{(1-(1+\delta)q^2)^2+(2\zeta q)^2}} = \frac{\delta}{1+\delta}, \tag{3.9}$$

one obtains that

$$q_\delta = \frac{\sqrt{2}}{2}\sqrt{\frac{1+2\delta}{\delta^2+\delta-2\zeta^2(1+2\delta)}}. \tag{3.10}$$

Note that q_δ is real if and only if $\zeta < \zeta_\delta = \sqrt{\frac{\delta^2+\delta}{2(1+2\delta)}}$. Since the transmissibility tends to an asymptote at the level of $\frac{\delta}{1+\delta}$ when q tends to ∞, ζ_δ is a critical value of ζ in the sense that: if $\zeta < \zeta_\delta$, there exists a finite q where the minimum of μ occurs; otherwise, μ is uniformly larger than $\frac{\delta}{1+\delta}$ and approaches $\frac{\delta}{1+\delta}$ when q tends to ∞. The curve with $\zeta = \zeta_\delta$ is shown in Fig. 3.5.

Note that q_p and q_b are the natural frequency and the anti-resonant frequency of the undamped case, respectively. For the damped case, the real natural frequency q_{pr} and anti-resonant frequency q_{br} for a specific damping ratio ζ can be obtained by setting the derivative of (3.5) to zero. Then, one obtains

$$q_{pr} = \sqrt{\frac{1+2\delta-\sqrt{1+8\zeta^2(1+2\delta)}}{2(\delta^2+\delta-2\zeta^2(1+2\delta))}}, \tag{3.11}$$

$$q_{br} = \sqrt{\frac{1+2\delta+\sqrt{1+8\zeta^2(1+2\delta)}}{2(\delta^2+\delta-2\zeta^2(1+2\delta))}}. \tag{3.12}$$

It is clear that if $\zeta \approx 0$, $q_{pr} \approx q_p$, and $q_{br} \approx q_b$ hold, but for a large ζ, it is not sufficient to use this estimation.

In summary, one obtains the following remarks:

Remark 1 *1. The parallel-connected inerter can effectively lower the invariant point that is independent of the damping ratio* ζ*;*
2. The magnitude at the natural frequency is a decreasing function with respect to both the damping ratio and the inertance-to-mass ratio; the magnitude at the anti-resonant frequency is an increasing function with respect to both the damping ratio and the inertance-to-mass ratio;
3. The isolation at high frequencies is weakened by using the parallel-connected inerter, where the magnitude tends to $\frac{\delta}{1+\delta}$ *when* q *tends to* ∞*.*

Analysis of C2

For this configuration, the transmissibility can be obtained as

$$\begin{aligned} \mu &= \frac{\mid \frac{kc}{b} - c\omega^2 + kj\omega \mid}{\mid \frac{kc}{b} - c\omega^2 - \frac{mc}{b}\omega^2 + (k - m\omega^2)j\omega \mid}, \\ &= \sqrt{\frac{\delta^2 q^2 + 4(1-\delta q^2)^2\zeta^2}{\delta^2(1-q^2)^2q^2 + 4(1-(1+\delta)q^2)^2\zeta^2}}. \end{aligned} \tag{3.13}$$

By rewriting (3.13) as

$$\mu = \sqrt{\frac{A\zeta^2 + B}{C\zeta^2 + D}},$$

where $A = 4(1-\delta q^2)^2$, $B = \delta^2 q^2$, $C = 4(1-(1+\delta)q^2)^2$, and $D = \delta^2(1-q^2)^2q^2$, the invariant points which are independent of damping can be similarly obtained by setting

$$\frac{A}{C} = \frac{B}{D},$$

that is,

$$\frac{1-\delta q^2}{1-(1+\delta)q^2} = \pm\frac{1}{1-q^2}.$$

For the case of plus sign, after simple calculation, one obtains $\delta q^4 = 0$, which leads to $q = 0$, a trivial result. For the case of minus sign, one obtains

$$\delta q^4 - 2(1+\delta)q^2 + 2 = 0.$$

Then, one can obtain the two nonzero invariant points as

$$q_{P,Q}^2 = 1 + \frac{1}{\delta} \pm \sqrt{1 + \frac{1}{\delta^2}}. \tag{3.14}$$

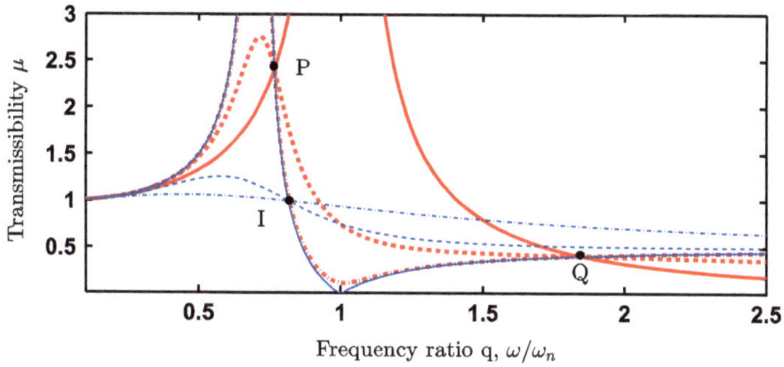

Fig. 3.6 Comparison of the transmissibilities of configurations $C1$ and $C2$ when $\delta = 1$. Red bold lines denote $C2$ and blue thin lines denote $C1$. The solid lines denote $\zeta = 0$; the dash lines denote $\zeta = \zeta_\delta = 0.5774$; the dash-dot lines denote $\zeta = \zeta_r = \sqrt{1+\delta} = \sqrt{2}$

Denote $q_P < q_Q$. It is easy to show that $q_P^2 < 1$ and $q_Q^2 > 2$, and both q_P and q_Q are decreasing functions with respect to δ. This indicates that, similar to the parallel-connected inerter, the series-connected inerter can also effectively lower the invariant points. Note that the magnitudes at P and Q are

$$\mu|_{q=q_P} = \left|\frac{1}{1-q_P^2}\right|, \quad \mu|_{q=q_Q} = \left|\frac{1}{1-q_Q^2}\right|.$$

Since $q_P^2 < 1$ and $q_Q^2 > 2$, one obtains

$$\mu|_{q=q_P} > 1 > \mu|_{q=q_Q}, \tag{3.15}$$

which means that for a finite δ, it is impossible to equalize the ordinates at the two invariant points.

A comparison of the transmissibilities of configurations $C1$ and $C2$ is shown in Fig. 3.6, where two invariant points P and Q of configuration $C2$ are depicted. It is shown that for the same damping ratio ζ, the behaviors of configurations $C1$ and $C2$ are totally different. For example, for the case of $\zeta = \zeta_r = \sqrt{2}$ (dash-dot lines in Fig. 3.6), $C1$ is overdamped while $C2$ behaves similarly to the undamped case of $C1$. This is caused by the series structure of $C2$, as by varying the damping ratio ζ from 0 to ∞, the configuration $C2$ is changed from the configuration with only a spring to the configuration with a parallel connection of a spring and an inerter.

In summary, one obtains the following remarks:

Remark 2 *1. Two invariant points, which are independent of the damping ratio, can be introduced by using the series-connected inerter, and both the two invariant points are decreasing functions with respect to the inertance-to-mass ratio;*

2. *For a finite inertance-to-mass ratio, the magnitude at the smaller invariant point is larger than* 1 *and the magnitude at the larger invariant point is smaller than* 1*;*

3. *The series arrangement $C2$ behaves between the configuration with only a spring and the configuration with a parallel connection of a spring and an inerter.*

3.4 H_∞ Optimization for Inerter-Based Isolators

In practice, in order to achieve good isolating performance, it is always desirable to minimize the maximum displacement of the object, which is known as H_∞ optimization (Cheung and Wong 2011a). In the previous section, it is shown that the invariant point, the resonant frequency, and the anti-resonant frequency are directly determined by the inertance-to-mass ratio δ. Therefore, in this section, H_∞ tuning procedures for a given δ will be proposed.

For the configuration $C1$ in Fig. 3.3, the optimal damping in H_∞ optimization for a given δ is ∞, which is a trivial solution, as in this case the object and the foundation are stiffly connected. For the configuration $C2$ with a given inertance-to-mass ratio δ, the optimal damping ratio ζ for the H_∞ performance is the one making the curve horizontally pass through the invariant P, as shown in Fig. 3.6. The rationality is based on the fixed-point theory (Den Hartog 1985, Sect. 3.3), the most favorable damping is the one making the curve horizontally pass through the highest invariant point. As demonstrated in Sect. 3.3, the magnitude of the invariant point P is always larger than that of the other invariant point Q. Therefore, based on this consideration, the optimal damping ratio ζ for configuration $C2$ can be obtained as follows:

Proposition 3.1 *For the configuration $C2$ with a given δ, the optimal damping ratio ζ in H_∞ optimization is*

$$\zeta_{opt} = \frac{1}{2}\sqrt{\delta(1 + \delta - \sqrt{1 + \delta^2})}. \tag{3.16}$$

Proof See Appendix.

Note that two invariant points can be introduced by using the series-connected inerter, and in order to further tune the two invariant points, an extra spring k_1 is incorporated. Then, three inerter-based isolators are proposed as shown in Fig. 3.8. The fixed-point theory (Den Hartog 1985, Sect. 3.3) is employed to derive the optimal parameters for these three inerter-based isolators. The fixed-point theory can be summarized as follows (Den Hartog 1985, Sect. 3.3).

Procedure 1 1. *For a given inertance-to-mass ratio δ, find the invariant points which are independent of the damping ratio ζ, and denote the two smaller invariant points as P and Q;*
2. *Adjust the spring stiffness ratio λ so that the ordinates at the invariant points P and Q are equal;*
3. *Calculate the damping ratio ζ_P and ζ_Q so that the curves of transmissibility μ vs. q horizontally pass through P and Q, respectively;*

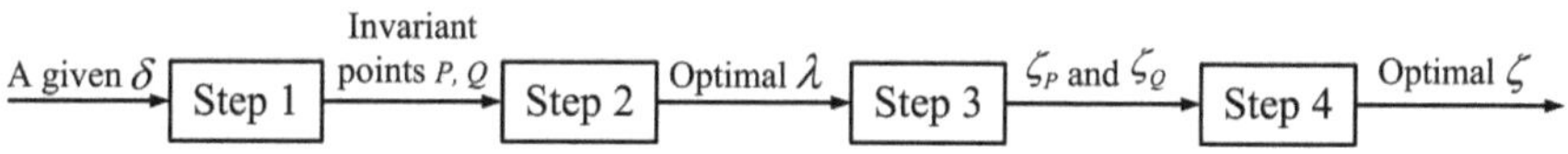

Fig. 3.7 Graphical representation of Procedure 1

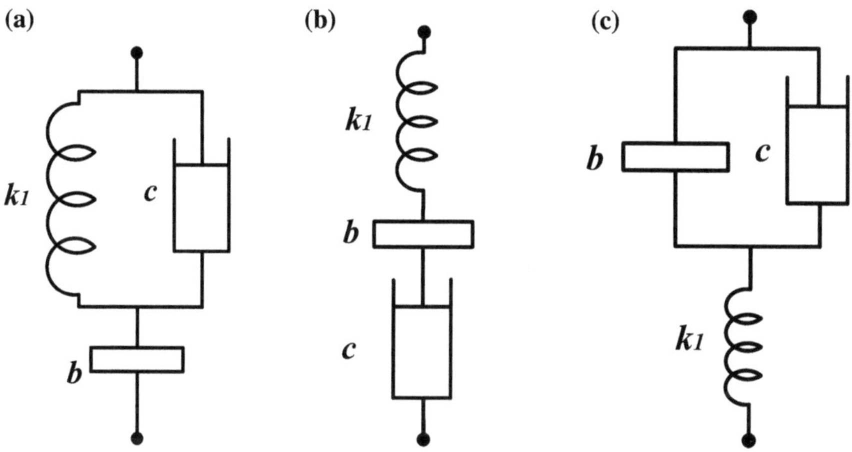

Fig. 3.8 Three configurations as $W(s)$ of the isolators in Fig. 3.1. **a** C3; **b** C4; **c** C5

4. *Obtain the optimal damping ratio as* $\zeta = \sqrt{\frac{\zeta_P^2+\zeta_Q^2}{2}}$.

A graphical representation of Procedure 1 is given in Fig. 3.7, indicating the required and output parameters in each step. According to this procedure, the optimal parameters λ and ζ for each configuration are derived subsequently (Fig. 3.8).

Remark 3 *The fixed-point theory (Den Hartog 1985, Sect. 3.3) actually yields a suboptimal but highly precise solution as demonstrated in (Nishihara and Asami 2002). The merit of the fixed-point theory is that an analytical solution can be easily derived, which makes it extensively employed in tuning dynamic vibration absorber (DVA) (or tuned mass damper (TMD)). See, for example, (Ren 2001; Cheung and Wong 2011a; Asami et al. 1991) and references therein. This is also the reason why it is employed in this paper. Please note that the optimal parameters derived in this section are "optimal" in the sense of the fixed-point theory using Procedure 1, which would be suboptimal in practice.*

Proposition 3.2 *The transmissibility for C3 can be obtained as*

$$\mu = \left| \frac{1 - \delta(1+\lambda)q^2 + 2\mathrm{j}\lambda(1-\delta q^2)q\zeta}{1 - (\delta + 1 + \delta\lambda)q^2 + \delta\lambda q^4 + 2\mathrm{j}\lambda(1-(1+\delta)q^2)q\zeta} \right|. \tag{3.17}$$

As shown in appendix, there are three invariant points for C3 which are denoted as P, Q, and R ($q_P < q_Q < q_R$), respectively. Following Procedure 1, the largest invariant point R can be derived as

$$q_R^2 = \frac{1}{\delta} + \frac{3}{2} + \sqrt{\left(\frac{1}{\delta} - \frac{3}{2}\right)^2 + \frac{4}{\delta}}, \tag{3.18}$$

which possesses a relatively large value ($q_R^2 \geq 3$). The optimal stiffness ratio λ can be obtained as

$$\lambda = \frac{2(q_R^4\delta(1+\delta) - (1+2\delta)q_R^2 + 1)}{\delta q_R^2(q_R^4\delta - 2(\delta+1)q_R^2 + 2)} \text{ or } \frac{2((1+2\delta)(1+\delta)q_R^2 - 2(1+\delta))}{q_R^2(\delta(1+2\delta)q_R^2 - 2(1+2\delta+2\delta^2))}. \tag{3.19}$$

The optimal damping ratio ζ can be obtained as

$$\zeta = \sqrt{\frac{\zeta_P^2 + \zeta_Q^2}{2}}, \tag{3.20}$$

where ζ_P^2 and ζ_Q^2 can be obtained as

$$\zeta_{P,Q}^2 = \left(\frac{1 - \delta(1+\lambda)q_{P,Q}^2}{1 - \delta q_{P,Q}^2}\right)\left(\frac{\delta(1+\lambda)(2 - (1+2\delta)q_{P,Q}^2) - (2\delta\lambda q_{P,Q}^2 - 1)(1 - \delta q_{P,Q}^2)}{4\lambda^2 q_{P,Q}^2}\right), \tag{3.21}$$

q_P^2 and q_Q^2 are solutions of the following quadratic function with respect to q^2:

$$q^4 - \left(\frac{2}{\delta\lambda}(1+\lambda+\delta+\lambda\delta) - q_R^2\right)q^2 + \frac{2}{\delta^2\lambda q_R^2} = 0. \tag{3.22}$$

Proof See appendix.

Procedure 2 *In summary, the H_∞ tuning procedure for C3 is given below:*

1. *Obtain q_R from (3.18);*
2. *Obtain λ_{opt} by substituting q_R into (3.19);*
3. *Obtain q_P and q_Q by solving (3.22);*
4. *Obtain ζ_p^2 and ζ_Q^2 by substituting q_P and q_Q into (3.21), respectively;*
5. *Obtain the optimal ζ_{opt} from (3.20).*

Note that in (Lazar et al. 2014), a similar tuning procedure was given for the configuration $C3$ by following the procedure given in (Den Hartog 1985) as well. The main difference between the method in this paper and the one in (Lazar et al. 2014) is the approach in calculating the optimal parameters λ and ζ. In this paper, the analytical solutions of the optimal λ and ζ are given, that is, (3.18), (3.19), and (3.21); while in (Lazar et al. 2014), the optimal λ and ζ are obtained relying on numerical iterations. Hence, the procedure in this paper is more convenient and reliable.

The transmissibility μ of $C3$ for $\delta = 0.2$ is illustrated in Fig. 3.9.

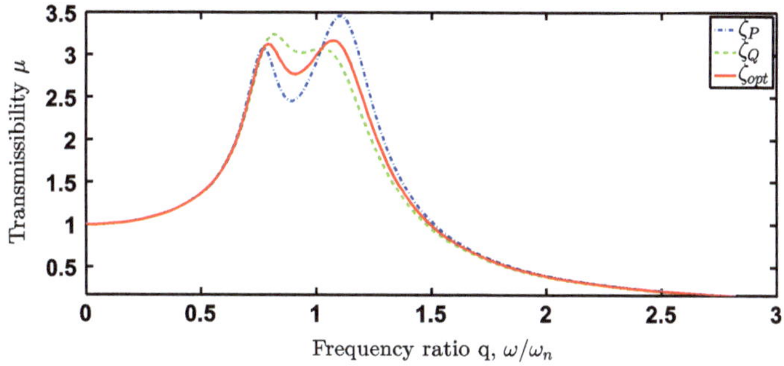

Fig. 3.9 Transmissibility μ for $C3$ when $\delta = 0.2$

Proposition 3.3 *The transmissibility for C4 can be obtained as*

$$\mu = \left| \frac{2(1-\delta(1+\lambda)q^2)\zeta + \mathrm{j}\delta q}{2(\delta\lambda q^4 - (1+\delta+\delta\lambda)q^2+1)\zeta + \mathrm{j}\delta(1-q^2)q} \right|. \tag{3.23}$$

Following Procedure 1, the optimal stiffness ratio λ can be obtained as

$$\lambda = \frac{1}{\delta}. \tag{3.24}$$

The optimal damping ratio ζ can be obtained as

$$\zeta_{opt} = \sqrt{\frac{\zeta_P^2 + \zeta_Q^2}{2}}, \tag{3.25}$$

where

$$\zeta_P^2 = \frac{\delta^2\left(1-\sqrt{\delta/(2+\delta)}\right)}{4\left((1+\delta)\sqrt{\delta/(2+\delta)}-\delta\right)\left((\delta+3)\sqrt{\delta/(2+\delta)}+\delta\right)}, \tag{3.26}$$

$$\zeta_Q^2 = \frac{\delta^2\left(1+\sqrt{\delta/(2+\delta)}\right)}{4\left((1+\delta)\sqrt{\delta/(2+\delta)}+\delta\right)\left((\delta+3)\sqrt{\delta/(2+\delta)}-\delta\right)}. \tag{3.27}$$

Proof See Appendix.

The transmissibility μ of $C4$ for $\delta = 0.2$ is illustrated in Fig. 3.10.

Proposition 3.4 *The transmissibility for C5 can be obtained as*

$$\mu = \left| \frac{1-\delta(1+\lambda)q^2 + \mathrm{j}2(\lambda+1)\zeta q}{1-(1+\delta+\delta\lambda)q^2+\delta\lambda q^4 + \mathrm{j}2\zeta(\lambda+1-\lambda q^2)q} \right|. \tag{3.28}$$

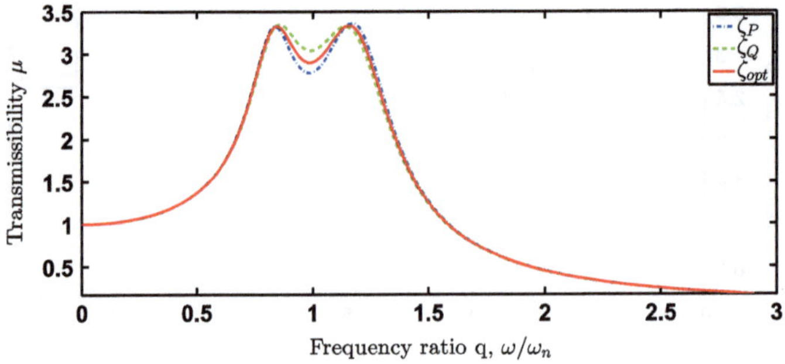

Fig. 3.10 Transmissibility μ for $C4$ when $\delta = 0.2$

Following Procedure 1, the optimal stiffness ratio λ can be obtained as

$$\lambda = \frac{1}{2\delta}\left(1 - 2\delta + \sqrt{1-2\delta}\right), \tag{3.29}$$

which requires $\delta < 1/2$. The optimal damping ratio ζ can be obtained as

$$\zeta_{opt} = \sqrt{\frac{\zeta_P^2 + \zeta_Q^2}{2}}, \tag{3.30}$$

where

$$\zeta_{P,Q}^2 = \frac{\left(1 - \delta(1+\lambda)q_{P,Q}^2\right)\left(1 + 2\delta + 2\delta\lambda - 3\delta\lambda q_{P,Q}^2\right)}{4(\lambda+1)\lambda q_{P,Q}^2} \tag{3.31}$$

and

$$q_{P,Q}^2 = \frac{1}{4\delta\lambda(\lambda+1)}\left(1 + 2\lambda + 2\delta(1+\lambda)^2 \pm \sqrt{\left(2\delta(1+\lambda)^2 + 1 - 2\lambda\right)^2 + 8\lambda}\right). \tag{3.32}$$

Proof See Appendix.

The transmissibility μ of $C5$ for $\delta = 0.2$ is illustrated in Fig. 3.11.

Comparison between the traditional DVA and the inerter-based isolators

Now, all the optimal parameters for these inerter-based isolators in H_∞ optimization have been derived. In this section, the performance of the inerter-based isolators will be compared with the traditional DVA as shown in Fig. 3.12. For the traditional DVA,

$$\mu = \sqrt{\frac{A\zeta^2 + B}{C\zeta^2 + D}},$$

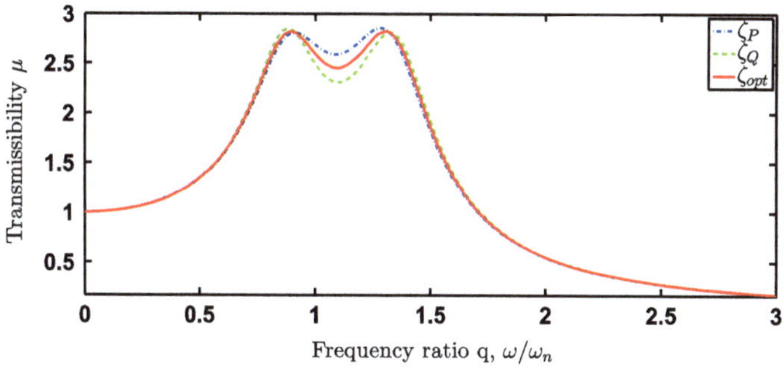

Fig. 3.11 Transmissibility μ for $C5$ when $\delta = 0.2$

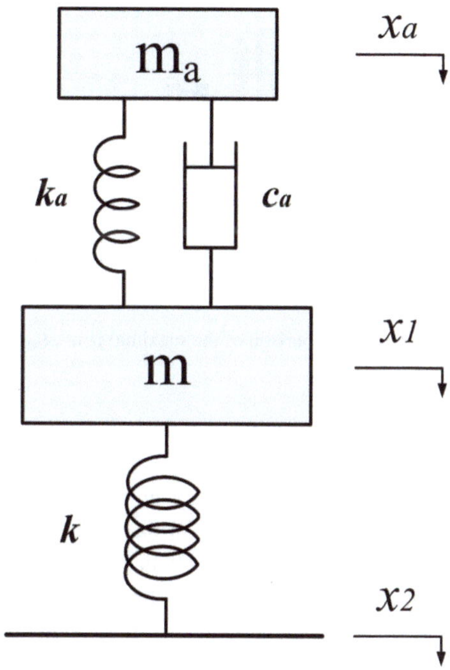

Fig. 3.12 The dynamic vibration absorber attached to the object mass

where $A = 4\lambda^2 q^2$, $B = (1 - \delta\lambda q^2)^2$, $C = 4\lambda^2\left(1 - (1+\delta)q^2\right)^2 q^2\zeta^2 + \left(1 - (1 + \delta + \delta\lambda)q^2 + \delta\lambda q^4\right)^2$, and the mass ratio δ and the stiffness ratio λ are defined as $\delta = \frac{m_a}{m}$ and $\lambda = \frac{k}{k_a}$, respectively.

It is well known that the optimal parameters for the traditional DVA (Ren 2001; Cheung and Wong 2011a; Asami et al. 1991) are

$$\lambda_{opt} = \frac{(\delta+1)^2}{\delta}, \ \zeta_{opt} = \frac{\delta}{1+\delta}\sqrt{\frac{3\delta}{8(1+\delta)}}.$$

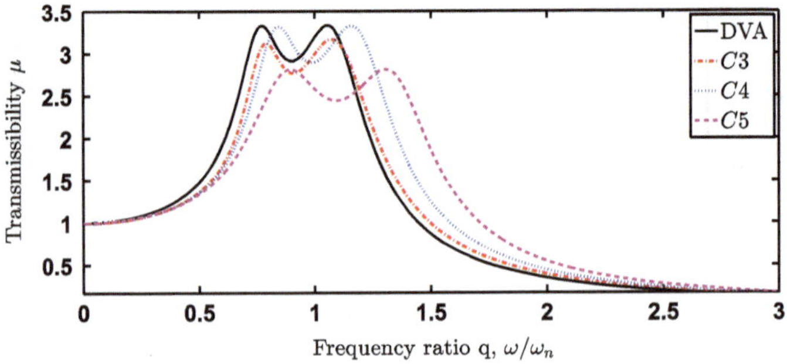

Fig. 3.13 Comparison between traditional DVA and inerter-based isolators when $\delta = 0.2$

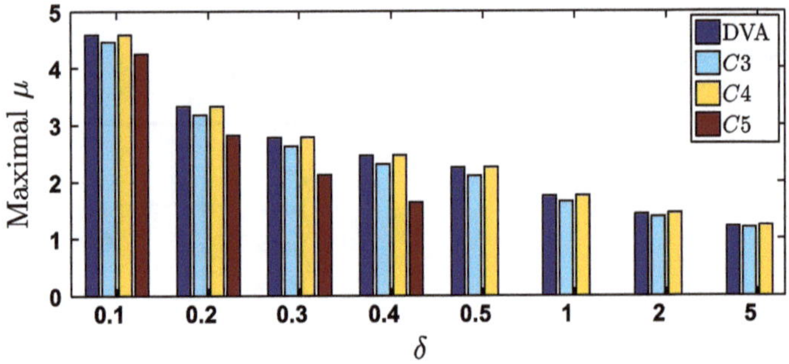

Fig. 3.14 Comparison of the maximal μ in H_∞ optimization

Figure 3.13 shows the comparison between the traditional DVA and the inerter-based isolators when the inertance-to-mass ratio (or mass ratio for traditional DVA) $\delta = 0.2$, where it is clearly shown that in terms of the same δ, the configuration $C4$ provides comparable performance compared with the traditional DVA, whereas both $C3$ and $C5$ perform better than the traditional DVA. Such an observation is confirmed by Fig. 3.14, where the comparison of the maximal μ with respect to different δ is shown. The comparison of the optimal stiffness ratio λ and damping ratio ζ with respect to different δ is shown in Fig. 3.15.

Note that the fundamental difference between the traditional DVA and the inerter-based isolators is that the inertance-to-mass ratio of the inerter-based isolators can easily be larger than the mass ratio of the traditional DVA, as large inertance can easily be obtained without increasing the physical mass of the whole system. For example, the inertance of a rack–pinion inerter or a ball–screw inerter can be significantly magnified by enlarging the gear ratios (Smith 2002; Chen et al. 2009). However, the mass ratio δ for the traditional DVA is practically less than 0.25 (Cheung and Wong 2011a; Inman 2008). From this point of view, the performance of the inerter-based isolators can be further improved compared with the traditional DVA, and the inerter-based isolators are potentially more attractive than the traditional DVA.

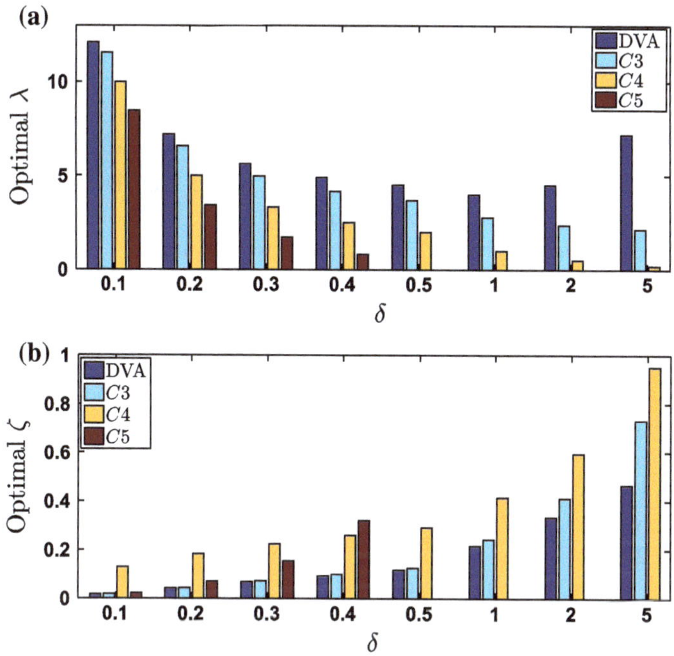

Fig. 3.15 Comparison of the optimal parameters in H_∞ optimization. **a** Optimal stiffness ratio λ; **b** optimal damping ratio ζ

3.5 H_2 Optimization for Inerter-Based Isolators

H_2 optimization aims to minimize the total vibration energy or the mean square motion of the object mass when white noise excitation is enforced (Cheung and Wong 2011b). In the case of random excitation such as wind loading instead of harmonic excitation, the H_2 optimization would be more practical than the H_∞ optimization. In this section, the analytical solutions for the inerter-based isolators in H_2 optimization will be derived and compared with the traditional DVA.

The performance measure to be minimized in H_2 optimization is defined as follows (Cheung and Wong 2011b; Asami et al. 1991):

$$I = \frac{E\left[x_1^2\right]}{2\pi S_0 \omega_n}, \tag{3.33}$$

where S_0 is the uniform power spectrum density function. Denoting $\mu = |H(\mathrm{j}q)|$, the mean square value of x_1 of the object mass m can be calculated as

$$E\left[x_1^2\right] = S_0 \int_{-\infty}^{\infty} |H(\mathrm{j}q)|^2 \,\mathrm{d}\omega = S_0 \omega_n \int_{-\infty}^{\infty} |H(\mathrm{j}q)|^2 \,\mathrm{d}q. \tag{3.34}$$

Substituting (4.12) into (4.11), one obtains

$$I = \frac{1}{2\pi} \int_{-\infty}^{\infty} |H(\mathrm{j}q)|^2 \, \mathrm{d}q, \tag{3.35}$$

which is exactly the definition of the H_2 norm of the transfer function $\hat{H}(s)$ by replacing jq in $H(\mathrm{j}q)$ with the Laplace variable s.

Therefore, the H_2 performance measure is rewritten as

$$I = \left\| \hat{H}(s) \right\|_2^2 . \tag{3.36}$$

In what follows, an analytical approach to calculating the H_2 norm of the transfer function $\hat{H}(s)$ will be presented according to (Doyle et al. 1992, Sect. 2.6), which has been used to derive analytical solutions for vehicle suspensions in (Hu et al. 2014; Scheibe and Smith 2009).

For a stable transfer function $\hat{H}(s)$, its H_2 norm can be calculated as (Doyle et al. 1992, Sect. 2.6)

$$\|\hat{H}(s)\|_2^2 = \|C(sI - A)^{-1}B\|_2^2 = CLC^T,$$

where A, B, and C are the minimal state-space realization $\hat{H}(s) = C(sI - A)^{-1}B$ and L is the unique solution of the Lyapunov equation given as

$$AL + LA^T + BB^T = 0. \tag{3.37}$$

We can write $\hat{H}(s)$ as

$$\hat{H}(s) = \frac{b_{n-1}s^{n-1} + \cdots + b_1 s + b_0}{s^n + a_{n-1}s^{n-1} + \cdots + a_1 s + a_0}$$

in its controllable canonical form below

$$\dot{x} = Ax + Bu, \;\; y = Cx,$$

where

$$A = \begin{bmatrix} 0 & 1 & 0 & \dots & 0 \\ 0 & 0 & 1 & \dots & 0 \\ \vdots & \vdots & \vdots & \vdots & \vdots \\ 0 & 0 & 0 & \dots & 1 \\ -a_0 & -a_1 & -a_2 & \dots & -a_{n-1} \end{bmatrix}, \; B = \begin{bmatrix} 0 \\ 0 \\ \vdots \\ 0 \\ 1 \end{bmatrix}, \; C = \begin{bmatrix} b_0, & b_1, & b_2 & \dots & b_{n-1} \end{bmatrix}.$$

Note that the analytical solution for the configuration $C1$ cannot be derived by using the above method, as the $\hat{H}(s)$ for $C1$ is not strictly proper. Actually, the H_2

norm of $\hat{H}(s)$ for $C1$ is infinity which can be obtained by observing Fig. 3.5, the area under the frequency response curve of $C1$ which represents the H_2 norm of the transfer function is infinity.

The procedure to derive the optimal parameters for $C2$, $C3$, $C4$, and $C5$ can be summarized as follows:

Procedure 3

1. *Analytically calculate the H_2 performance measure I using the method discussed above. Denote the performance measure as $I = F(\lambda)\zeta + \frac{G(\lambda)}{\zeta}$, where $F(\lambda)$ and $G(\lambda)$ are functions of λ with $F(\lambda) > 0$, $G(\lambda) > 0$;*
2. *Obtain the equations of optimal ζ and I as $\zeta_{opt} = \sqrt{\frac{F(\lambda)}{G(\lambda)}}$ and $I_{opt} = 2\sqrt{F(\lambda)G(\lambda)}$, respectively;*
3. *Obtain the optimal λ as the one minimizing $F(\lambda)G(\lambda)$, denoted as λ_{opt};*
4. *Obtain the optimal ζ and I by substituting λ_{opt} into the equations obtained in Step 2, respectively.*

Note that in Step 1 of Procedure 3, it includes the case that $F(\lambda)$ and $G(\lambda)$ are constants with respect to λ. Following Procedure 3, the optimal parameters for $C2$, $C3$, $C4$, and $C5$ in the H_2 optimization will be derived subsequently.

Proposition 3.5 *For the configuration $C2$, the H_2 performance measure in (4.13) is*

$$I_{c2} = \frac{1-\delta+\delta^2}{\delta^2}\zeta + \frac{1}{4\zeta}. \tag{3.38}$$

For a given δ, the optimal ζ is

$$\zeta_{opt} = \frac{\delta}{2\sqrt{1-\delta+\delta^2}}.$$

After substituting ζ_{opt} into (4.23), the optimal I_{c2} is

$$I_{c2,opt} = \frac{\sqrt{1-\delta+\delta^2}}{2\delta}.$$

Proof Equation (4.23) can be obtained by direct calculation, and then the optimal ζ and $I_{c2,opt}$ can be obtained subsequently.

Proposition 3.6 *For the configuration $C3$, the H_2 performance measure in (4.13) is*

$$I_{c3} = \frac{1-\delta+\delta^2}{\delta^2}\zeta + \frac{1-2\delta\lambda+\delta^2\lambda^2+\delta^2\lambda}{4\lambda^2\delta^2\zeta}. \tag{3.39}$$

For a given δ, the optimal λ can be obtained as

$$\lambda_{opt} = \begin{cases} \frac{2}{\delta(2-\delta)}, & \delta < 2, \\ \infty, & \delta \geq 2. \end{cases}$$

Note that in the case of $\delta \geq 2$, C3 reduces to C2. For a given δ and λ, the optimal ζ can be obtained as

$$\zeta_{opt} = \frac{1}{2\lambda}\sqrt{\frac{1 - 2\delta\lambda + \delta^2\lambda}{1 - \delta + \delta^2}}.$$

Then, the optimal I_{c3} can be obtained by substituting ζ_{opt} and λ_{opt} into (4.40).

Proof Equation (4.40) can be obtained by direct calculation. The optimal λ can be obtained by checking the second part in (4.40). Since both parts in (4.40) are positive, the optimal ζ can be obtained subsequently.

Proposition 3.7 *For the configuration C4, the H_2 performance measure in (4.13) is*

$$I_{c4} = \frac{1 - 2\delta\lambda + \delta^2\lambda^2 + 2\delta^2\lambda - \delta + \delta^2}{\delta^2}\zeta + \frac{1}{4\zeta}. \tag{3.40}$$

For a given δ, the optimal λ can be obtained as

$$\lambda_{opt} = \begin{cases} \frac{1-\delta}{\delta}, & \delta < 1, \\ 0, & \delta \geq 1. \end{cases}$$

Note that in the case of $\delta \geq 1$, C4 reduces to C2. For a given δ and λ, the optimal ζ can be obtained as

$$\zeta_{opt} = \frac{1}{2}\sqrt{\frac{\delta^2}{1 - 2\delta\lambda + \delta^2\lambda^2 + 2\delta^2\lambda - \delta + \delta^2}}.$$

Then, the optimal I_{c4} can be obtained by substituting ζ_{opt} and λ_{opt} into (3.40).

Proof The proof is omitted as it is similar to that of Proposition 3.6.

Proposition 3.8 *For the configuration C5, the H_2 performance measure in (4.13) is*

$$I_{c5} = (\lambda + 1)^2\zeta + \frac{\delta^3\lambda^3 + \delta(3\delta - 2)\lambda^2 + (1 - 2\delta + 3\delta^3)\lambda + \delta^2}{4\lambda\zeta}. \tag{3.41}$$

For a given δ and λ, the optimal ζ and I_{c5} can be obtained as

$$\zeta_{opt} = \frac{1}{2(1+\lambda)}\sqrt{\frac{\delta^3\lambda^3 + \delta(3\delta - 2)\lambda^2 + (1 - 2\delta + 3\delta^3)\lambda + \delta^2}{\lambda}}, \tag{3.42}$$

$$I_{c5,opt} = (\lambda + 1)\sqrt{\frac{\delta^3\lambda^3 + \delta(3\delta - 2)\lambda^2 + (1 - 2\delta + 3\delta^3)\lambda + \delta^2}{\lambda}}. \tag{3.43}$$

Let $\mathcal{Q}$ be the set of real, positive solutions λ of the quartic equation

$$4\delta^2\lambda^4 + (11\delta - 6)\delta\lambda^3 + (2 - 6\delta + 9\delta^2)\lambda^2 + \delta^2\lambda - \delta^2 = 0. \tag{3.44}$$

The optimal λ is chosen from the elements of $\mathcal{Q}$ as well as 0 that makes $I_{c5,opt}$ minimum. If the optimal λ is 0, configuration $C5$ reduces to $C1$.

Proof Equation (4.41) can be obtained by direct calculation. Since both parts in (4.41) are positive, the optimal ζ and I_{c5} can be obtained as in (3.42) and (3.43), respectively in a straightforward manner. In terms of (3.43), by making the derivative of $I_{c5,opt}$ with respect to λ zero, the quartic equation (3.44) can be obtained, and then the optimal λ can be selected from the real, positive solutions of the quartic equation as well as ∞.

Comparison between the traditional DVA and the inerter-based isolators

Now, all the optimal parameters for the inerter-based isolators in H_2 optimization have been derived. In this section, the performance of these inerter-based isolators will be compared with the traditional DVA as shown in Fig. 3.12.

For the traditional DVA shown in Fig. 3.12, the H_2 performance measure can be derived as

$$I_{DVA} = \frac{1+\delta}{\delta}\zeta + \frac{(\delta+1)^2 - \delta(\delta+2)\lambda + \delta^2\lambda^2}{4\lambda^2\delta^2\zeta}, \tag{3.45}$$

where the mass ratio δ and the stiffness ratio λ are defined as $\delta = m_a/m$ and $\lambda = k/k_a$.

Similar to the inerter-based isolators, the optimal parameters can be obtained as

$$\lambda_{opt} = \frac{2(\delta+1)^2}{\delta(\delta+2)},$$

$$\zeta_{opt} = 4\sqrt{\frac{\delta^3(3\delta+4)}{(\delta+1)^3}},$$

$$I_{DVA,opt} = \frac{1}{2}\sqrt{\frac{3\delta+4}{\delta(\delta+1)}}.$$

Figures 3.16, 3.17, and 3.18 show the comparison between the traditional DVA and the inerter-based isolators in H_2 optimization. As shown in Fig. 3.16, for the same δ, the inerter-based isolators $C5$ and $C3$ perform better than the traditional DVA when δ is less than 0.44 and 1.2, respectively, and the configuration $C3$ performs slightly worse than the traditional DVA. As shown in Fig. 3.16, when $\delta < 0.44$, the configuration $C5$ performs best among all the inerter-based isolators. From Fig. 3.17, it is shown that the damping ratios ζ of the inerter-based isolators are normally smaller than the traditional DVA. The detailed values of the parameters are given in Table 3.2, where it is shown that when $\delta = 0.2$, the inerter-based isolator $C3$ and $C5$ can provide 8.75% and 49.06% improvement compared with the traditional DVA.

Similar to the H_∞ optimization, the fundamental difference between the traditional DVA and the inerter-based isolators is that relatively large value of inertance

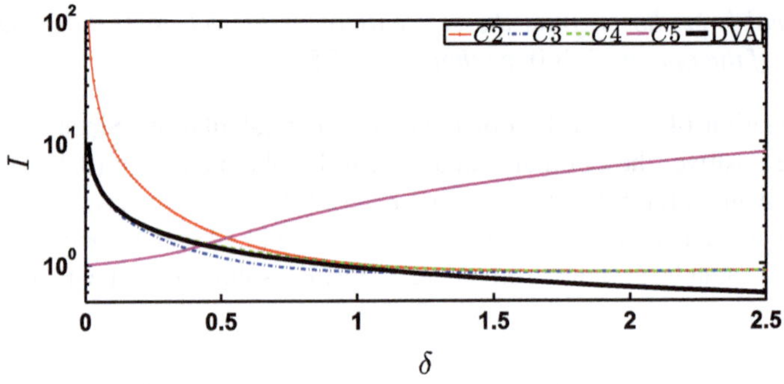

Fig. 3.16 Comparison between traditional DVA and inerter-based isolators in H_2 optimization

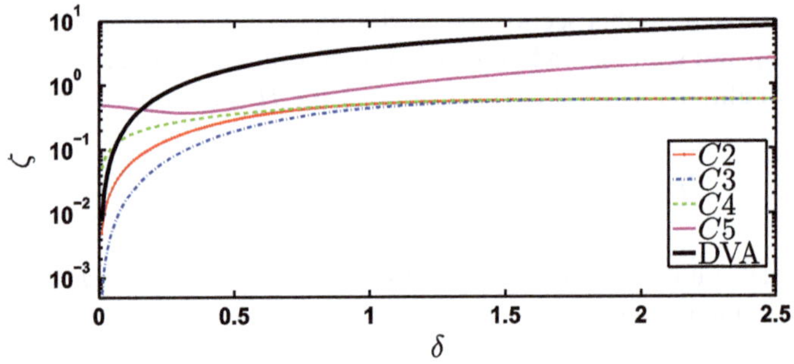

Fig. 3.17 Optimal damping ratio ζ in H_2 optimization

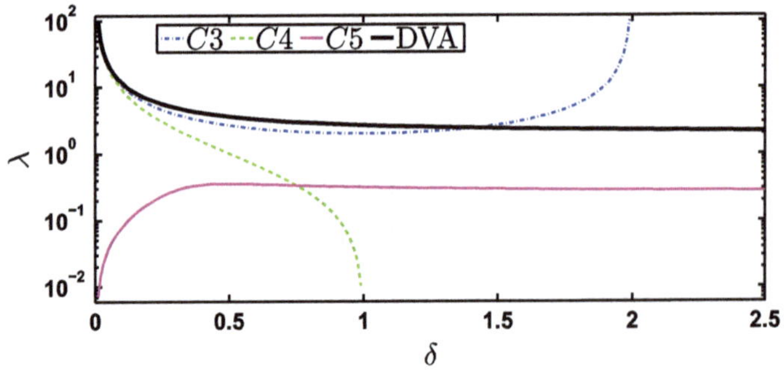

Fig. 3.18 Optimal stiffness ratio λ in H_2 optimization

can easily be achieved without increasing the physical mass of the isolation system (Smith 2002; Chen et al. 2009), whereas the attached mass m_a is normally quite small and the typical mass ratio δ for the traditional DVA is less than 0.25 (Cheung and Wong 2011a; Inman 2008). In this sense, the performance of the inerter-based isolators can be further improved by increasing the inertance-to-mass ratio δ even $\delta > 0.25$, which is a potential advantage of the inerter-based isolators compared with the traditional DVA.

Table 3.2 Comparison of optimal parameters in H_2 optimization

(a) H_2 performance measure I

δ	DVA	$C2$	$C3$	$C4$	$C5$
0.1	3.1261	9.5394	2.9787	3.1623	1.0479
0.2	2.1890	4.5826	1.9975	2.2361	1.1152
0.3	1.7723	2.9627	1.5607	1.8257	1.2184
0.4	1.5236	2.1794	1.3077	1.5811	1.3798
0.5	1.3540	1.7321	1.1456	1.4142	1.6015
1	0.9354	1.0000	0.8660	1.0000	3.1087
2	0.6455	0.8660	0.8660	0.8660	6.5065
5	0.3979	0.9165	0.9165	0.9165	16.9393

(b) Optimal stiffness ratio λ

δ	DVA	$C3$	$C4$	$C5$
0.1	11.5238	10.5263	9.0000	0.0796
0.2	6.5455	5.5556	4.0000	0.1787
0.3	4.8986	3.9216	2.3333	0.2824
0.4	4.0833	3.1250	1.5000	0.3426
0.5	3.6000	2.6667	1.0000	0.3542
1	2.6667	2.0000	0	0.3139
2	2.2500	∞	0	0.2815
5	2.0571	∞	0	0.2623

(c) Optimal damping ratio ζ

δ	DVA	$C2$	$C3$	$C4$	$C5$
0.1	0.2274	0.0524	0.0164	0.1581	0.4495
0.2	0.5837	0.1091	0.0476	0.2236	0.4014
0.3	0.9816	0.1688	0.0889	0.2739	0.3704
0.4	1.3930	0.2294	0.1376	0.3162	0.3827
0.5	1.8053	0.2887	0.1909	0.3536	0.4367
1	3.7417	0.5000	0.4330	0.5000	0.9004
2	6.8853	0.5774	0.5774	0.5774	1.9810
5	13.2637	0.5455	0.5455	0.5455	5.3157

3.6 Conclusions

In this paper, the performance of inerter-based isolators has been investigated by applying five configurations with inerter in a "uni-axial" isolation system. In the first part of this paper, the frequency responses of the inerter in parallel connection and the one in series connection are analyzed. It has been analytically demonstrated that both the parallel-connected inerter and the series-connected inerter one can effectively lower the invariant points, and the isolation for high frequencies can be weakened by using inerter. In the second part of this paper, both H_∞ and H_2 performances have been considered for the proposed inerter-based isolators. The fixed-point theory and the analytical method in calculating H_2 norm are employed to analytically derive the optimal parameters in H_∞ and H_2 optimization, respectively. The performances of the inerter-based isolators have also been compared with the traditional DVA to show the benefits of the inerter-based isolators. On one hand, it has been shown that for the same mass ratio or inertance-to-mass ratio, two inerter-based isolators perform better than the traditional DVA. On the other hand, two unique properties make the inerter-based isolators potentially more attractive than the traditional DVA: first, a large inertance can easily be obtained for inerter without increasing the physical mass of the whole system; second, the inerter is a built-in element and there is no need to mount an additional mass to the object to be isolated.

In practical applications of the inerter-based isolators, the large transmission ratios employed in the physical embodiments of inerter will amplify the internal friction of the rotating device with a gain that is equal to the square of the transmission ratio. This could lead to an amount of damping at a system level larger than the optimal one, which may render the proposed inerter-based isolators far from an ideal design. More research work needs to be carried to find low-friction designs to be used with high amplification ratio.

Appendix

Proof of Proposition 3.1

Observing Fig. 3.6, it is shown that the curve horizontally passing through P indicates the optimal damping. This optimal damping can be obtained by solving the following equation:

$$\left.\frac{\partial \mu^2}{\partial q^2}\right|_{q=q_P} = 0. \tag{3.46}$$

Denote $\mu = \sqrt{\frac{n}{m}}$, where $n = \delta^2 q^2 + 4(1-\delta q^2)^2\zeta^2$, $m = \delta^2(1-q^2)^2 q^2 + 4(1-(1+\delta)q^2)^2\zeta^2$. Equation (3.46) can be written in another form as

$$n'm - m'n = 0,$$

where $n' = \partial n/\partial q^2$ and $m' = \partial m/\partial q^2$. For the invariant point P,

$$\frac{n}{m} = \frac{1}{(1-q^2)^2} = \frac{(1-\delta q^2)^2}{(1-(1+\delta)q^2)^2},$$

therefore,

$$(1-q^2)^2 n' - m' = 0.$$

Since

$$n' = -8(1 - \delta q^2)\delta\zeta^2 + \delta^2,$$

$$m' = -8(1 - (1 + \delta)q^2)(\delta + 1)\zeta^2 + \delta^2(1 - q^2)(1 - 3q^2),$$

after substituting q_P into (3.14), one obtains

$$\zeta_{opt} = \frac{1}{2}\sqrt{\delta(1 + \delta - \sqrt{1 + \delta^2})}.$$

Proof of Proposition 3.2

Denote

$$A = 4\lambda^2(1 - \delta q^2)^2 q^2, \ B = (1 - \delta(1 + \lambda)q^2)^2,$$

$$C = 4\lambda^2(1 - (1 + \delta)q^2)^2 q^2, \ D = (1 - (\delta + 1 + \delta\lambda)q^2 + \delta\lambda q^4)^2.$$

Then, μ in (3.17) can be rewritten as

$$\mu = \sqrt{\frac{A\zeta^2 + B}{C\zeta^2 + D}}. \tag{3.47}$$

To find the invariant points which are independent of damping, it requires

$$\frac{A}{C} = \frac{B}{D},$$

that is,

$$\frac{1 - \delta q^2}{1 - (1 + \delta)q^2} = \pm\frac{1 - \delta(1 + \lambda)q^2}{1 - (\delta + 1 + \delta\lambda)q^2 + \delta\lambda q^4}.$$

With the plus sign, after cross-multiplication, one obtains $\delta^2\lambda q^6 = 0$, which leads to the trivial solution $q = 0$. With the minus sign, after simple calculation, one obtains

$$\delta^2\lambda q^6 - 2\delta(\lambda + \delta + 1 + \delta\lambda)q^4 + 2(2\delta + 1 + \delta\lambda)q^2 - 2 = 0, \tag{3.48}$$

which is a cubic form in q^2. Therefore, there are three invariant points for the configuration $C3$.

Denoting these three invariant points as P, Q, and R ($q_P < q_Q < q_R$), separately, one obtains

$$q_P^2 + q_Q^2 + q_R^2 = \frac{2}{\delta\lambda}(\lambda + \delta + 1 + \lambda\delta), \tag{3.49}$$

$$q_P^2 q_Q^2 q_R^2 = \frac{2}{\delta^2\lambda}, \tag{3.50}$$

$$q_P^2 q_Q^2 + q_P^2 q_R^2 + q_Q^2 q_R^2 = \frac{2}{\delta^2\lambda}(2\delta + 1 + \delta\lambda). \tag{3.51}$$

Since at points P and Q, the values of μ are independent of ζ, then in the case of $\zeta = \infty$, one obtains

$$\left|\frac{1 - \delta q_P^2}{1 - (1+\delta)q_P^2}\right| = \left|\frac{1 - \delta q_Q^2}{1 - (1+\delta)q_Q^2}\right|.$$

It can be checked that

$$\frac{1 - \delta q_P^2}{1 - (1+\delta)q_P^2} > 0, \quad \frac{1 - \delta q_Q^2}{1 - (1+\delta)q_Q^2} < 0.$$

Then, one obtains

$$\frac{1 - \delta q_P^2}{1 - (1+\delta)q_P^2} = -\frac{1 - \delta q_Q^2}{1 - (1+\delta)q_Q^2}.$$

After cross-multiplication and simplification, one obtains

$$2\delta(1+\delta)q_P^2 q_Q^2 - (q_P^2 + q_Q^2)(1 + 2\delta) + 2 = 0. \tag{3.52}$$

Substituting (3.50) and (3.51) into (3.52), one can obtain a quadratic equation with respect to q_R^2 as

$$\delta\lambda(1 + 2\delta)q_R^4 - 2(\lambda + 2\delta\lambda + 3\delta + 2\delta^2 + 1 + 2\lambda\delta^2)q_R^2 + 4(1+\delta) = 0. \tag{3.53}$$

Note that q_R is the same solution as both (3.48) and (3.53) for the same δ and λ. Solving λ from (3.48) and (3.53), separately, one obtains

$$\lambda = \frac{2(q_R^4\delta(1+\delta) - (1+2\delta)q_R^2 + 1)}{\delta q_R^2(q_R^4\delta - 2(\delta+1)q_R^2 + 2)}, \tag{3.54}$$

$$\lambda = \frac{2((1+2\delta)(1+\delta)q_R^2 - 2(1+\delta))}{q_R^2(\delta(1+2\delta)q_R^2 - 2(1 + 2\delta + 2\delta^2))}. \tag{3.55}$$

Equating the solutions and simplifying the results, one obtains

$$\delta q_R^4 - (2 + 3\delta)q_R^2 + 2 = 0. \tag{3.56}$$

Then, one obtains q_R^2 as shown in (3.18).

From (3.18), it is easy to show that $q_R^2 \geq 3$, which is relatively large compared with the natural frequency. This can explain why only invariant points P and Q are involved in the H_∞ tuning of $C3$.

In this way, the optimal λ can be obtained by substituting q_R^2 in (3.18) into (3.54) or (3.55). After obtaining λ, all the three invariant points can be obtained by solving

$$q^4 - \left(\frac{2}{\delta\lambda}(1+\lambda+\delta+\lambda\delta) - q_R^2\right) q^2 + \frac{2}{\delta^2\lambda q_R^2} = 0,$$

which is obtained from (3.50) and (3.51).

The procedure of calculating the optimal damping ratio ζ is similar to the procedure in appendix, where the optimal ζ makes the gradients at invariant points P and Q zero. After calculation and simplification, one obtains (3.21). Taking an average of ζ_P^2 and ζ_Q^2, one obtains the optimal ζ_{opt} as in (3.20).

Proof of Proposition 3.3

Denote

$$A = 4(1 - \delta(1+\lambda)q^2)^2,\ B = \delta^2 q^2,$$

$$C = 4(1 - (1+\delta+\delta\lambda)q^2 + \delta\lambda q^4)^2,\ D = \delta^2(1-q^2)^2 q^2,$$

and μ in (3.23) can be rewritten as

$$\mu = \sqrt{\frac{A\zeta^2 + B}{C\zeta^2 + D}}. \tag{3.57}$$

To find the invariant points which are independent of damping, it requires

$$\frac{A}{C} = \frac{B}{D},$$

that is,

$$\frac{1 - \delta(1+\lambda)q^2}{1 - (1+\delta+\delta\lambda)q^2 + \delta\lambda q^4} = \pm\frac{1}{1-q^2}.$$

Again, with the plus sign, one obtains the trivial solution zero, and with the minus sign, one obtains

$$\delta(1+2\lambda)q^4 - 2(1+\delta+\delta\lambda)q^2 + 2 = 0. \tag{3.58}$$

Then, one obtains the two invariant points P and Q ($q_P < q_Q$) as

$$q_{P,Q}^2 = \frac{1+\delta+\delta\lambda \pm \sqrt{(1+\delta+\delta\lambda)^2 - 2\delta(1+2\lambda)}}{\delta(1+2\lambda)}. \tag{3.59}$$

Letting the ordinates at invariant points P and Q equal, one has

$$\left|\frac{1}{1-q_P^2}\right| = \left|\frac{1}{1-q_Q^2}\right|.$$

It can be checked that $\frac{1}{1-q_P^2} > 0$ and $\frac{1}{1-q_Q^2} < 0$. Then, one obtains

$$\frac{1}{1-q_P^2} = -\frac{1}{1-q_Q^2}.$$

After cross-multiplication and simplification, one has

$$q_P^2 + q_Q^2 = 2. \tag{3.60}$$

Considering (3.58), one obtains

$$\frac{2(1+\delta+\delta\lambda)}{\delta(1+2\lambda)} = 2,$$

which leads to (3.24).

Similar to the method in appendix, the optimal ζ can be obtained by making μ to have zero gradients at invariant points P and Q. After calculation and simplification, one obtains

$$\zeta_{P,Q}^2 = \frac{q_{P,Q}^2\delta^2}{4\left(1-\delta(1+\lambda)q_{P,Q}^2\right)\left(1+2\delta+2\delta\lambda-\delta(1+3\lambda)q_{P,Q}^2\right)}.$$

After substituting (3.59) and (3.24), one obtains (3.26) and (3.27).

Taking an average of ζ_p^2 and ζ_Q^2, one obtains the optimal ζ_{opt} as in (3.25).

Proof of Proposition 3.4

Denote

$$A = 4(\lambda+1)^2q^2,\ B = (1-\delta(1+\lambda)q^2)^2,$$

$$C = 4(\lambda+1-\lambda q^2)^2q^2,\ D = (1-(1+\delta+\delta\lambda)q^2+\lambda\delta q^4)^2.$$

Then, μ in (3.28) can be rewritten as

$$\mu = \sqrt{\frac{A\zeta^2+B}{C\zeta^2+D}}. \tag{3.61}$$

To find the invariant points which are independent of damping, it requires

$$\frac{A}{C} = \frac{B}{D},$$

that is,

$$\frac{\lambda + 1}{\lambda + 1 - \lambda q^2} = \pm \frac{1 - \delta(1 + \lambda)q^2}{1 - (1 + \delta + \delta\lambda)q^2 + \delta\lambda q^4}.$$

Similarly, with plus sign, one obtains the trivial solution zero, and with minus sign, one obtains

$$2\delta\lambda(\lambda + 1)q^4 - \left(1 + 2\lambda + 2\delta(1 + \lambda)^2\right) q^2 + 2(\lambda + 1) = 0. \tag{3.62}$$

Thus, one obtains the two invariant points P and Q ($q_P < q_Q$) as in (3.32).

Letting the ordinates at invariant points P and Q equal, one has

$$\left|\frac{\lambda + 1}{\lambda + 1 - \lambda q_P^2}\right| = \left|\frac{\lambda + 1}{\lambda + 1 - \lambda q_Q^2}\right|.$$

It can be checked that $\frac{\lambda+1}{\lambda+1-\lambda q_P^2} > 0$ and $\frac{\lambda+1}{\lambda+1-\lambda q_Q^2} < 0$. Then, one obtains

$$\frac{\lambda + 1}{\lambda + 1 - \lambda q_P^2} = -\frac{\lambda + 1}{\lambda + 1 - \lambda q_Q^2}.$$

After cross-multiplication and simplification, one has

$$q_P^2 + q_Q^2 = \frac{2(\lambda + 1)}{\lambda}.$$

Comparing with (3.62), one obtains

$$\frac{1 + 2\lambda + 2\delta(1 + \lambda)^2}{2\delta\lambda(\lambda + 1)} = \frac{2(\lambda + 1)}{\lambda},$$

which leads to

$$2\delta\lambda^2 - 2(1 - 2\delta)\lambda + 2\delta - 1 = 0.$$

It can be checked that this equation has real solutions if and only if

$$\delta \leq 1/2.$$

Under this condition, the optimal λ can be obtained as in (3.29).

Note that if $\delta = \frac{1}{2}$, from (3.29), one has $\lambda = 0$ or $k = \infty$. In this case, $C5$ reduces to $C1$. Thus, the more reasonable assumption is $\delta < \frac{1}{2}$ rather than $\delta \leq \frac{1}{2}$.

Similarly, the optimal ζ can be obtained by making μ to have zero gradients at invariant points P and Q. After calculation and simplification, one obtains ζ_P^2 and ζ_Q^2 as in (3.31).

Taking an average of ζ_p^2 and ζ_Q^2, one obtains the optimal ζ_{opt} as in (3.30).

References

Asami, T., Wakasono, T., Kameoka, K., Hasegawa, M., & Sekiguchi, H. (1991). Optimum design of dynamic absorbers for a system subjected to random excitation. *JSME International Journal Series III*, *34*(2), 218–226.

Carrella, A., Brennan, M. J., Waters, T. P., & Lopes, V, Jr. (2012). Force and displacment transimissibility of a nonlinear isolator with high-static-low-dynamic-stiffness. *International Journal of Mechanical Sciences*, *55*(1), 22–29.

Chen, M. Z. Q., Papageorgiou, C., Scheibe, F., Wang, F. C., & Smith, M. C. (2009). The missing mechanical circuit element. *IEEE Circuits and Systems Magazine*, *9*(1), 10–26.

Chen, M. Z. Q., Hu, Y., Li, C., & Chen, G. (2015). Performance benefits of using inerter in semiactive suspensions. *IEEE Transactions on Control System Technology*, *23*(4), 1571–1577.

Chen, M. Z. Q., Hu, Y., Huang, L., & Chen, G. (2014). Influence of inerter on natural frequencies of vibration systems. *Journal of Sound and Vibration*, *333*(7), 1874–1887.

Cheung, Y. L., & Wong, W. O. (2011a). H-infinity optimization of a variant design of the dynamic vibration absorber-Revisited and new results. *Journal of Sound and Vibration*, *330*(16), 3901–3912.

Cheung, Y. L., & Wong, W. O. (2011b). H_2 optimization of a non-traditional dynamic vibration abosorber for vibration control of structures under random force excitation. *Journal of Sound and Vibration*, *330*(6), 1039–1044.

Den Hartog, J. P. (1985). *Mechanical Vibrations*. New York: Dover Publications, INC.

Doyle, J. C., Francis, B. A., & Tannenbaum, A. R., et al. (1992). *Feedback Control Theory*. Oxford: Maxwell Macmillan Int.

Dylejko, P. G., & MacGillivray, I. R. (2014). On the concept of a transmission absorber to suppress internal resonance. *Journal of Sound and Vibration*, *333*, 2719–2734.

Hu, Y., Chen, M. Z. Q., & Shu, Z. (2014). Passive vehicle suspensions employing inerters with multiple performance requirements. *Journal of Sound and Vibration*, *333*(8), 2212–2225.

Hu, Y., Chen, M. Z. Q., Shu, Z., & Huang, L. (2014). Vibration analysis for isolation system with inerter. In *Proceedings of the 33rd Chinese Control Conference* (pp. 6687–6692). China: Nanjing.

Inman, D. J. (2008). *Engineering Vibration* (3rd ed.). Upper Saddle River: Prentice-Hall Inc.

Lazar, I. F., Neild, S. A., & Wagg, D. J. (2014). Using an inerter-based device for structural vibration suppression. *Earthquake Engineering and Structure Dynamics*, *43*(8), 1129–1147.

Marian, L., & Giaralis, A. (2014). Optimal design of a novel tuned mass-damper-inerter (TMDI) passive vibration control configuration for stochastically support-excited structural systems. *Probabilistic Engineering Mechanics*, *38*, 156–164.

Nishihara, O., & Asami, T. (2002). Closed-form solutions to the exact optimizations of dynamic vibration absorbers (minimizations of the maximum amplitude magnification factors). *Journal of Vibration and Acoustics*, *124*(4), 576–582.

Ren, M. Z. (2001). A variant design of the dynamic vibration absorber. *Journal of Sound and Vibration*, *245*(4), 762–770.

Rivin, E. I. (2003). *Passive Vibration Isolation*. New York: ASME Press.

Piersol, A. G., & Paez, T. L. (2010). *Harris' Shock and Vibration Handbook* (6th ed.). New York: McGraw-Hill.

Scheibe, F., & Smith, M. C. (2009). Analytical solutions for optimal ride comfort and tyre grip for passive vehicle suspensions. *Vehile System Dynamics*, *47*(10), 1229–1252.

Smith, M. C. (2002). Synthesis of mechanical networks: The inerter. *IEEE Transaction on Automatic Control*, *47*(1), 1648–1662.

Smith, M. C., & Wang, F. C. (2004). Performance benefits in passive vehicle suspensions employing inerters. *Vehicle System Dynamics*, *42*(4), 235–257.

Wang, F.-C., & Chan, H. A. (2011). Vehicle suspensions with a mechatronic network strut. *Vehicle System Dynamics*, *49*(5), 811–830.

Wang, F.-C., Hsieh, M.-R., & Chen, H.-J. (2012). Stability and performance analysis of a full-train system with inerters. *Vehicle System Dynamics*, *50*(4), 545–571.

Wang, K., Chen, M. Z. Q., & Hu, Y. (2014). Synthesis of biquadratic impedances with at most four passive elements. *Journal of the Franklin Institute*, *351*(3), 1251–1267.

Chapter 4
Inerter-Based Dynamic Vibration Absorption System

Abstract This chapter is concerned with the H_∞ and H_2 optimization problem for inerter-based dynamic vibration absorbers (IDVAs). The proposed IDVAs are obtained by replacing the damper in the traditional dynamic vibration absorber (TDVA) with some inerter-based mechanical networks. It is demonstrated in this chapter that adding one inerter alone to the TDVA provides no benefits for the H_∞ performance and negligible improvement (less than 0.32% improvement over the TDVA when the mass ratio less than 1) for the H_2 performance. This implies the necessity of introducing another degree of freedom (element) together with inerter to the TDVA. Therefore, four different IDVAs are proposed by adding an inerter together with a spring to the TDVA, and significant improvement for both the H_∞ and H_2 performances is obtained. Numerical simulations in dimensionless form show that more than 20 and 10% improvement can be obtained for the H_∞ and H_2 performances, respectively. Besides, for the H_∞ performance, the effective frequency band can be further widened by using inerter.

Keywords Dynamic vibration absorber · IDVA · H_∞ optimization · H_2 optimization · Dimensionless analysis

4.1 Introduction

Dynamic vibration absorber (DVA) is an auxiliary mass system attached to a vibrating primary system to reduce undesired vibration, which is widely used in the fields of civil and mechanical engineering for its simple design and high reliability (Den Hartog 1985). In the first DVA proposed by Frahm in 1909 (Frahm 1909), only a spring was employed, and it was useful only in a narrow band of frequency. In 1928, the damping mechanism was introduced by Ormondroyd and Den Hartog (1928), which is a parallel arrangement of a spring and a damper, and as a result, the effective frequency band was significantly widened. It was also pointed out in Ormondroyd and Den Hartog (1928) that for the spring–damper DVA (in this chapter, it is called the traditional DVA or TDVA) and undamped primary system, there were two frequencies called fixed points, where the magnitudes were independent of the damping, and the

M. Z. Q. Chen and Y. Hu, *Inerter and Its Application in Vibration Control Systems*, https://doi.org/10.1007/978-981-10-7089-1_4

optimal setting of the spring stiffness was the one equalizing the magnitudes at the fixed points, and the optimal damping was the one making the curves of the frequency response horizontally pass through the fixed points. Such a tuning method is still in use today and currently known as the fixed-point method (Den Hartog 1985), which has been demonstrated to be a suboptimal H_∞ optimization method (Nishihara and Asami 2002). The exact solutions were analytically derived in Nishihara and Asami (2002) and it was also shown that the fixed-point method actually yielded an approximate but highly precise solution (with less than 0.5% deviation when the mass ratio less than 1). Another common performance measure of tuning DVA is the H_2 performance measure, which is desirable when the primary system subjected to random excitations. The objective of H_2 optimization is to optimize the total vibration energy of the system over all frequencies (Crandall and Mark 1963). For the TDVA with undamped primary systems, the optimal tuning frequency and damping ratio were investigated in Crandall and Mark (1963), and then the analytical solutions were derived in Asami et al. (1991). For damped primary systems, various design methods and tuning criteria have been proposed, such as those in Anh and Nguyen (2013), Asami et al. (2002), Ghosh and Basu (2007), Bekdas and Nigdeli (2013), and the applications of the TDVA in nonlinear and distributed primary systems have been investigated (Cheung and Wong 2009; Pai and Schulz 2000; Miguelez et al. 2010). The active DVAs utilizing feedback control actions have also been proposed (Gao et al. 2013; Si et al. 2014; Zhan et al. 2013).

Vibration absorption is one of the potential applications of inerter (Smith 2002). In Smith (2002), the problem of designing inerter-based networks to absorb vibration at a specific frequency was studied. Thereafter, the suppression of vibration over a broadband frequency by using inerter has been proposed. In Lazar et al. (2014), an inerter-based configuration ($C4$ in this chapter) was employed between adjacent storeys to suppress the vibration of a multistorey building. In Hu et al. (2015), optimal solutions for several inerter-based isolators (including all the configurations except $C5$ in this paper) were algebraically derived based on a "uniaxial" vibration isolation system. In Marian and Giaralis (2014), a new configuration incorporating an inerter was proposed and applied to a mechanical cascaded (chain-like) systems. In Brzeski et al. (2014), the dynamics of a tuned mass absorber with an additional viscous damper and an inerter attached to the pendulum was investigated.

In this chapter, a novel structure for inerter-based DVAs (IDVAs) is proposed by replacing the damper in the TDVA with some inerter-based mechanical networks, and both the H_∞ and H_2 performances of the proposed IDVAs are investigated. It is demonstrated in this chapter that adding an inerter alone to the TDVA, no matter it is in parallel connection or in series connection, provides no benefits for the H_∞ performance and negligible benefits (less than 0.32% improvement over the TDVA when the mass ratio less than 1) for the H_2 performance. In contrast, by adding an inerter together with a spring to the TDVA (e.g. $C3$, $C4$, $C5$, and $C6$ in this chapter), both H_∞ and H_2 performances can be significantly improved. Over 20% improvement compared with the TDVA can be obtained for the H_∞ performance, and the effective frequency band can also be further widened by using inerter. For the H_2 performance, it is analytically demonstrated that the IDVAs proposed in this chapter perform surely better than the TDVA and over 10% improvement is obtained in numerical simulation. Moreover, a minmax framework directly using the resonance frequencies is proposed for the H_∞ optimization, and an algebraic method to analytically calculate the H_2 norm is employed for the H_2 optimization. All these constitute the main contributions of this chapter.

4.2 Preliminary

The traditional spring–damper DVA is shown in Fig. 4.2a, where the mass M is the primary mass, i.e., the main structure the vibration of which is to be controlled. The spring–damper–mass (k, c, m) system is the DVA to be designed. The commonly used method for parameter tuning is the so-called fixed-point method (Den Hartog 1985), which can be summarized as follows.

The frequency response of the spring–damper DVA with respect to various values of absorber damping is shown in Fig. 4.1. It is obvious that if the damping is zero, the spring–damper DVA reduces the spring-only DVA (Frahm 1909); while is the damping is ∞, the two masses are rigidly connected together then a single-degree-of-freedom system is obtain. For both cases, the magnitudes are infinity, as shown in Fig. 4.1. Therefore, there must exist a value of damping where the peak of the frequency response is minimal. This result can also can be explained by from the energy dissipation point of view. The amplitudes of the masses are reduced by converting the kinetic energy into heat via the damper (Den Hartog 1985). The work done by the damping force can be calculated by the force times the relative displacement. For the case of zero damping, no work is done, and hence the amplitude is infinity; for the case of infinity damping, the two masses are clamped together such that the relative displacement is zero, and hence no work is done either. There must exist a damping where the work done by the damping force is maximal and then the amplitudes are minimized.

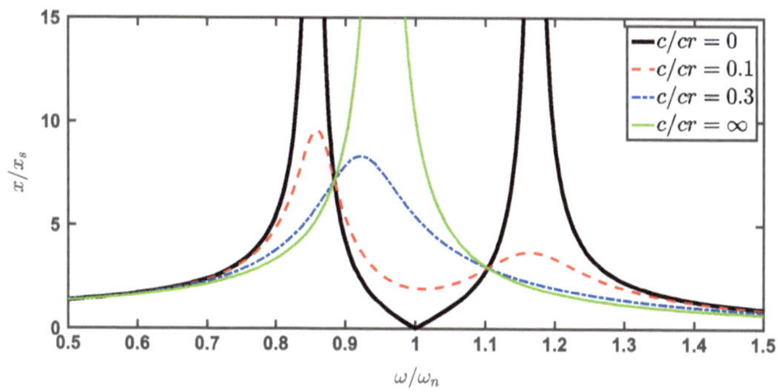

Fig. 4.1 Frequency response of the primary mass with respect to various values of absorber damping for traditional spring–damper DVA (Den Hartog 1985)

Observing Fig. 4.1, it is shown that two invariant points independent of the damping are depicted. Therefore, the most favorable curve is the one which has equal heights of the invariant points and a horizontal tangent through these invariant points (Den Hartog 1985). Then, two steps are generally required for the fixed-point method: first, a proper choice of the spring stiffness where the heights of the two invariant points are equal; second, a proper choice of the damping coefficient where the curve passes through the invariant points horizontally. Since it normally not possible to find a damping coefficient such that the curve simultaneously passes through the two invariant points horizontally, some approximations are usually employed (Den Hartog 1985).

4.3 Inerter-Based Dynamic Vibration Absorbers

Figure 4.2 shows the comparison between the IDVAs proposed in this chapter and the TDVA, where the IDVA is obtained by replacing the damper in the TDVA with some inerter-based mechanical networks. The entire networks employed in this chapter are shown in Fig. 4.3. The equations of motion for the whole system in the Laplace domain are

$$Ms^2x = F + F_d - Kx, \tag{4.1}$$

$$ms^2x_a = -F_d, \tag{4.2}$$

$$F_d = (k + sY(s))(x_a - x), \tag{4.3}$$

where $Y(s)$ is the admittance of the inerter-based passive mechanical networks and F_d is the force of the DVA imposed on the primary mass M.

From (4.2) and (4.3), one obtains,

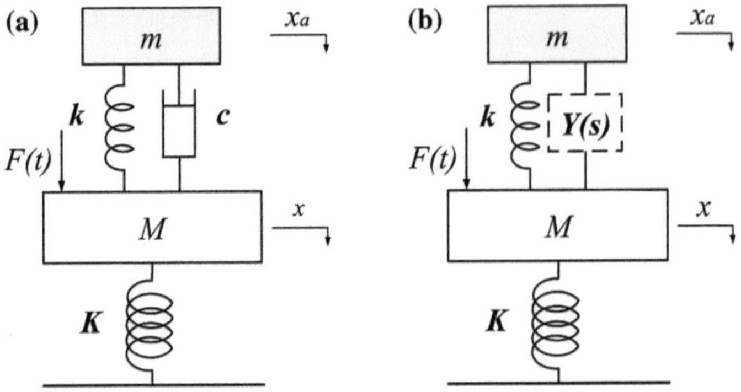

Fig. 4.2 Dynamic vibration absorbers (DVA): **a** traditional dynamic vibration absorber (TDVA); **b** inerter-based dynamic vibration absorber (IDVA)

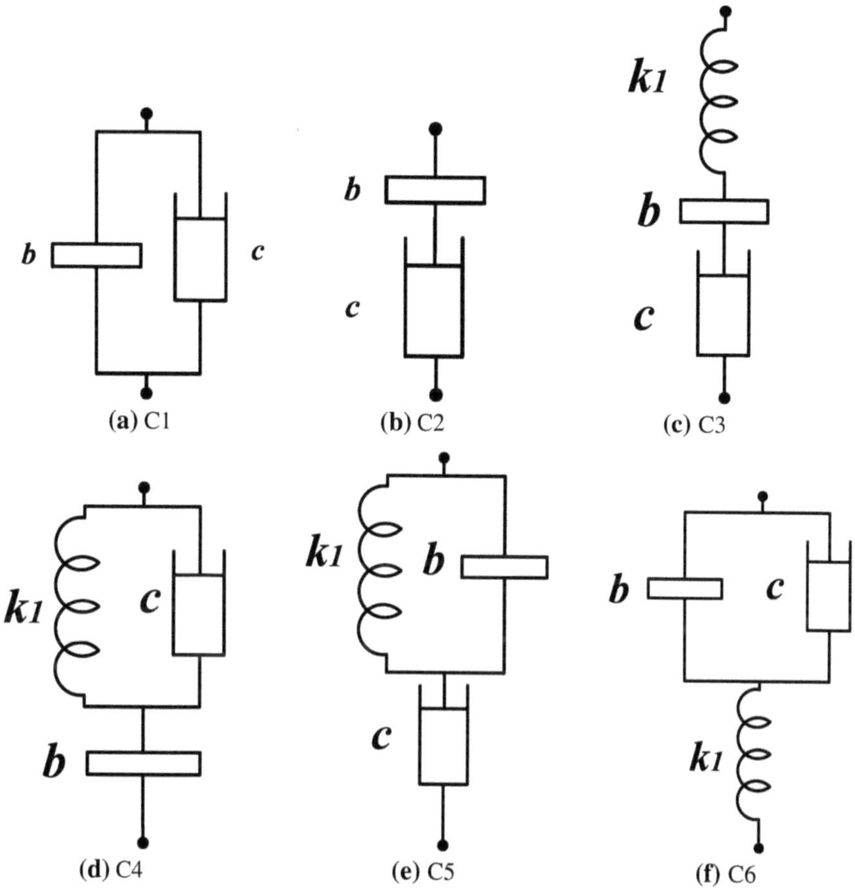

Fig. 4.3 The employed inerter-based networks as $Y(s)$ in Fig. 4.2

$$F_d = -R(s)x,$$

where

$$R(s) = \frac{(k + sY(s))ms^2}{k + ms^2 + sY(s)}.$$

Then, one obtains the displacement transfer function as

$$H(s) = \frac{x}{x_s} = \frac{1}{\frac{s^2}{\omega_n^2} + \frac{1}{K}R(s) + 1}, \tag{4.4}$$

where $x_s = F/K$ and $\omega_n = \sqrt{\frac{K}{M}}$ are the static displacement and natural frequency of the primary system, respectively.

The admittance of each network in Fig. 4.3 is shown in Table 4.1, where $Y_i(s)$, $i = 1, \ldots, 6$ corresponds to C_i, $i = 1, \ldots, 6$ in Fig. 4.3, respectively. Substituting each $Y_i(s)$ into (4.4), one can obtain the detailed transfer function for each configuration. To obtain the dimensionless representation of each configuration, the following dimensionless parameters are defined as

$$\left.\begin{aligned} \mu &= \tfrac{m}{M} : \text{ mass ratio} \\ \delta &= \tfrac{b}{m} : \text{ inertance-to-mass ratio} \\ \zeta &= \tfrac{c}{2\sqrt{mk}} : \text{ damping ratio} \\ \eta &= \tfrac{\omega_b}{\omega_m} : \text{ corner frequency ratio} \\ \gamma &= \tfrac{\omega_m}{\omega_n} : \text{ natural frequency ratio} \\ \lambda &= \tfrac{\omega}{\omega_n} : \text{ forced frequency ratio} \end{aligned}\right\} \tag{4.5}$$

where

$$\left.\begin{aligned} \omega_m &= \sqrt{\tfrac{k}{m}} : \text{ natural frequency of the DVA} \\ \omega_b &= \sqrt{\tfrac{k_1}{b}} : \text{ corner frequency of the DVA} \\ \omega_n &= \sqrt{\tfrac{K}{M}} : \text{ natural frequency of the primary system} \end{aligned}\right\} \tag{4.6}$$

Remark 4 *In this chapter, the force–current analogy between mechanical and electrical networks is employed, and admittance is defined to be the ratio of force to velocity, which agrees with the usual electrical terminology (see Smith 2002 for details). Such a definition is consistent with some books (Shearer and Murphy 1967, p. 328), but not others which use the force–voltage analogy (Hixson 1988).*

Remark 5 *Since the natural frequencies would be perturbed by using inerter as demonstrated in Chen et al. (2014), ω_m and ω_n are not the real natural frequencies of the whole system. Neither is ω_b the real corner frequency. Here, these notations are employed just for dimensionless representations.*

Replacing s with $j\omega$ in (4.4), the frequency response functions in a dimensionless form can be obtained as

$$H_i(j\lambda) = \frac{R_{ni} + jI_{ni}}{R_{mi} + jI_{mi}}, \ i = 1, \ldots, 6, \tag{4.7}$$

where R_{ni}, I_{ni}, R_{mi}, and I_{mi}, $i = 1, \ldots, 6$ are functions with respect to λ, γ, δ, and ζ. The detailed representations are given in Appendix.

Table 4.1 Admittance $Y(s)$ for each configuration in Fig. 4.3

$Y_1(s) = bs + c$	$Y_2(s) = \frac{1}{\frac{1}{bs}+\frac{1}{c}}$	$Y_3(s) = \frac{1}{\frac{s}{k_1}+\frac{1}{c}+\frac{1}{bs}}$
$Y_4(s) = \frac{1}{\frac{1}{\frac{k_1}{s}+c}+\frac{1}{bs}}$	$Y_5(s) = \frac{1}{\frac{1}{\frac{k_1}{s}+bs}+\frac{1}{c}}$	$Y_6(s) = \frac{1}{\frac{1}{bs+c}+\frac{s}{k_1}}$

4.4 H_∞ Optimization for the IDVAs

4.4.1 Minmax Optimization Problem Formulation

The objective of the H_∞ optimization is to minimize the maximum magnitude of the frequency response $|H_i(j\lambda)|$, $i = 1, \ldots, 6$, which is known as the H_∞ norm of $H_i(s)$ with $s = j\lambda$. For the TDVA, the fixed-point method (Den Hartog 1985) is commonly used to analytically obtain the optimal parameters (Den Hartog 1985, Sect. 3.3). Since there always exist more than two fixed points with respect to the damping ratio for IDVAs, it is difficult to obtain simple and analytical representations for optimal parameters. Given this fact, in this chapter, a minmax optimization problem is formulated as follows to directly minimize the magnitude at resonance frequencies.

For a given mass ratio μ, solving the follow minmax problem

$$\min_{\delta,\gamma,\eta,\zeta}\left(\max_{\lambda_l}\left(|H_i(j\lambda_l)|\right)\right), i = 1, \ldots, 6 \tag{4.8}$$

subject to $\delta \geq 0$, $\gamma \geq 0$, $\eta \geq 0$, $\zeta \geq 0$, and λ_l, $l = 1, \ldots, N$, are the real and positive solutions of the following equation:

$$\frac{\partial |H_i(j\lambda)|^2}{\partial \lambda^2} = 0, \tag{4.9}$$

where $i = 1, \ldots, 6$ corresponds to the six IDVAs in Fig. 4.3, respectively.

The underlying idea of the minmax problem (4.8) and (4.9) is, instead of using the fixed points to approximately minimize the H_∞ norm as done in the fixed-point method (Den Hartog 1985), here the resonance frequencies are directly used to exactly minimize the H_∞ norm. This is inspired by the method in Nishihara and Asami (2002), where the two resonance frequencies were employed to derive the exact solutions for the TDVA. Note that the solution set of (4.9), that is λ_l, $l = 1, \ldots, N$, contains the resonance frequencies, anti-resonance frequencies, and other frequencies where the curves horizontally pass through. Since the largest magnitude of the frequency response, representing the H_∞ norm of the transfer function, only occurs at resonance frequencies, it is sufficient to minimize $\max_{\lambda_l}(|H_i(j\lambda_l)|)$, $l = 1, \ldots, N$, to obtain the optimal H_∞ norm of the transfer function $H_i(s)$.

Equation (4.9) can be transformed into a polynomial function with respect to λ^2 as follows. From (4.7), $|H_i(j\lambda)|^2$ can be written as

$$|H_i(j\lambda)|^2 = \frac{n}{m},$$

where $n = R_{ni}^2 + I_{ni}^2$, $m = R_{mi}^2 + I_{mi}^2$. Since

$$\frac{\partial |H_i(j\lambda)|^2}{\partial \lambda^2} = \frac{n'm - m'n}{m^2},$$

where $n' = \frac{\partial n}{\partial \lambda^2}$ and $m' = \frac{\partial m}{\partial \lambda^2}$, (4.9) is equivalent to

$$n'm - m'n = 0, \tag{4.10}$$

which is an equation of λ^2 with different orders for different configurations.

Problem (4.8) and (4.10) is a constrained optimization problem, and the equality constraint (4.10) can be transformed into the objective function by employing $\lambda_l = f(\delta, \gamma, \eta, \zeta)$. In this chapter, a direct search method is employed to solve the constrained optimization problem (4.8) and (4.10) by using the Matlab solver *patternsearch* with multiple starting points.

4.4.2 Comparison Between the TDVA and IDVAs

For the TDVA, the optimal parameters can be analytically obtained as (Den Hartog 1985):

$$\gamma_{opt} = \sqrt{\frac{1}{1+\mu}}, \quad \zeta_{opt} = \sqrt{\frac{3\mu}{8(1+\mu)}},$$

and the optimal height at the two fixed points are $\sqrt{\frac{2+\mu}{\mu}}$.

4.4.2.1 Performance Limitation of $C1$ and $C2$

In this subsection, it will be demonstrated that configurations $C1$ and $C2$ provide no improvement for the H_∞ performance compared with the TDVA.

For configuration $C1$, by directly using the fixed-point method in Den Hartog (1985), the optimal parameters for $C1$ can be analytically obtained as

$$\gamma_{opt} = \frac{\sqrt{1+(1+\mu)\delta}}{1+\mu}, \quad \zeta_{opt} = \sqrt{\frac{3\mu}{8(1+\mu)}},$$

and the optimal height at the two fixed points is $\sqrt{\frac{2+\mu+2\delta(1+\mu)}{\mu}}$. It is obvious that the optimal δ is 0, which means that the parallel inerter in configuration $C1$ provides no

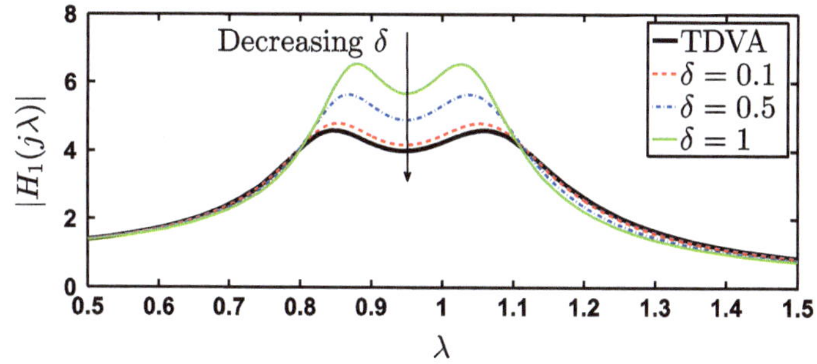

Fig. 4.4 Comparison between the TDVA and $C1$ when $\mu = 0.1$ with different δ

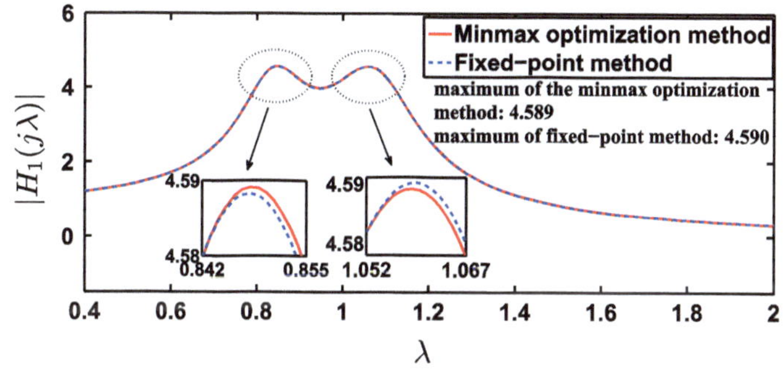

Fig. 4.5 Comparison between the minmax optimization method in this chapter and the fixed-point method when $\mu = 0.1$

improvement in the H_∞ optimization. Such an observation is shown in Fig. 4.4 with $\mu = 0.1$.

The minmax optimization method proposed in this chapter is also applicable for $C1$ and a comparison between the method in this chapter and the fixed-point method is shown in Fig. 4.5. As shown in Fig. 4.5, the results by these two methods highly coincide with each other and the results are consistent with the analytical solutions in Nishihara and Asami (2002, Table 2), which demonstrates the effectiveness of the method in this chapter.

In what follows, it will be shown that for configuration $C2$, the series-connected inerter provides no improvement for the H_∞ performance as well. To show the influence of δ, the problem (4.8) is slightly modified as: for a given μ and δ,

$$\min_{\gamma,\zeta}\left(\max_{\lambda_l}\left(|H_2(j\lambda_l)|\right)\right),$$

subject to $\gamma \geq 0$, $\eta \geq 0$, $\zeta \geq 0$, and λ_l, $l = 1, \ldots, N$, are the real and positive solutions of (4.10). Figure 4.6 shows the comparison between $C2$ with different δ and the TDVA when $\mu = 0.1$, where it is clearly shown that the maximum of $|H_2(j\lambda)|$ is decreased by increasing δ and if δ is sufficiently large, the frequency response of

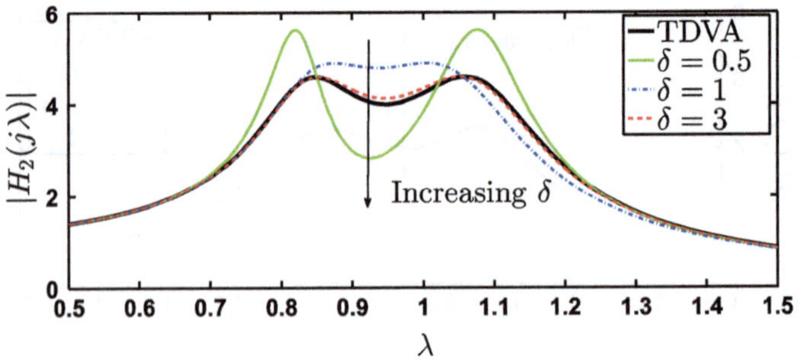

Fig. 4.6 Comparison between the TDVA and $C2$ when $\mu = 0.1$ with different δ

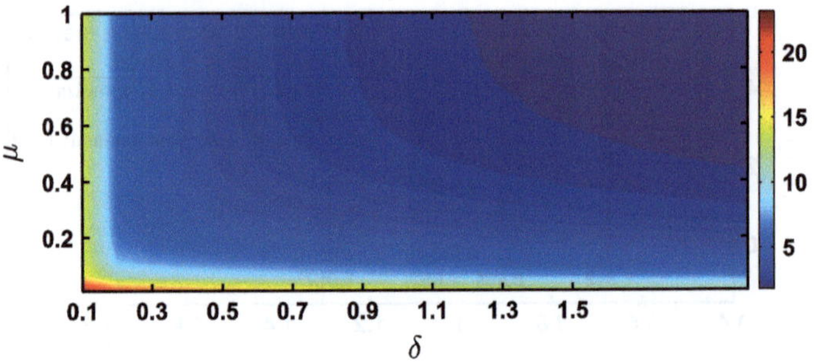

Fig. 4.7 $\max(|H_2(j\lambda)|)$ with different μ and δ

$C2$ coincides with that of the TDVA. Such an observation is also confirmed by other choices of μ, as shown in Fig. 4.7. Therefore, it is sufficient to conclude that for a single series arrangement of an inerter and a damper, the series inerter provides no improvement for the H_∞ performance of the isolation system.

The IDVAs $C1$ and $C2$ represent the two ways of adding an inerter to the TDVA, that is, the parallel connection ($C1$) and the series connection ($C2$). Now, it has been demonstrated that adding a single inerter alone to the TDVA, no matter it is in parallel connection or in series connection, provides no improvement for the H_∞ performance. Therefore, other degrees of freedom should be introduced, which is the motivation of introducing IDVAs $C3$, $C4$, $C5$, and $C6$ by adding an inerter together with a spring to the TDVA.

4.4.2.2 Performance Benefits of $C3$, $C4$, $C5$, and $C6$

In this subsection, it will be shown that after adding another degree of freedom, that is the spring k_1, the H_∞ performance will be significantly improved compared with the TDVA.

The optimization problem (4.8) with the constraint (4.10) is solved for configurations $C3$, $C4$, $C5$, and $C6$, separately, where a ninth-order polynomial of equation

Table 4.2 Maximum magnitude max $|H(j\lambda)|$ in the H_∞ optimization

μ	TDVA (Nishihara and Asami 2002)	$C3$	$C4$	$C5$	$C6$
0.01	14.1796	11.0330	11.0860	12.9216	11.0351
0.02	10.0530	7.8340	7.9064	9.1498	7.8352
0.05	6.4080	5.0159	5.1194	5.8051	5.0210
0.1	4.5892	3.6175	3.7448	4.1379	3.6208
0.2	3.3254	2.6552	2.7986	2.9877	2.6616
0.5	2.2480	1.8513	1.9941	2.0198	1.8521
1	1.7457	1.4893	1.6127	1.5809	1.4893
2	1.4279	1.2697	1.3629	1.3157	1.2697
5	1.1942	1.1166	1.1702	1.1766	1.1166
10	1.1033	1.0602	1.0918	1.0934	1.0603

(4.10) with respect to λ^2 is obtained. The exact solutions of the TDVA in Nishihara and Asami (2002) are employed for comparison and the detailed parameter values are shown in Tables 4.2, 4.3, and 4.4. Table 4.2 shows that all the IDVAs $C3$, $C4$, $C5$, and $C6$ can improve the H_∞ performance compared with the TDVA, where $C3$ performs the best and the order of the performance is $C3 > C6 > C4 > C5 > TDVA$ ("$>$" means performing better) with an exception for $\mu >= 1$. However, since the mass ratio is normally quite small and practically less than 0.25 (Inman 2008; Cheung and Wong 2011b), it is sufficient to conclude that $C3 > C6 > C4 > C5 > TDVA$. Such a conclusion is also confirmed by Fig. 4.8, where the comparison of the IDVAs over the TDVA in the range of $0 < \mu \leq 0.25$ is shown. As shown in the right figure of Fig. 4.8, 8 to 26% improvement can be obtained for the IDVAs. The other parameters in the range of $0 < \mu \leq 0.25$ are depicted in Fig. 4.9. It should be noted that although the optimal γ and ζ for $C3$ are almost identical to the TDVA, as shown in Table 4.3 and Fig. 4.9, over 22% improvement can be provided by $C3$ compared with the TDVA. Moreover, the spring k_1 is better to be in series connection for the H_∞ performance, given the fact that $C3$ and $C6$ are superior to $C4$ and $C5$.

The frequency responses of the IDVAs and the TDVA when $\mu = 0.1$ are shown in Fig. 4.10, where one sees that the magnitudes of the IDVAs around 1 are much flatter than those of the TDVA, and the effective frequency band is much larger than that of the TDVA.

Table 4.3 Optimal natural frequency ratio γ and damping ratio ζ in the H_∞ optimization

μ	TDVA (Nishihara and Asami 2002)	$C3$	$C4$	$C5$	$C6$
(a) Optimal natural frequency ratio γ					
0.01	0.9902	0.9900	0.9957	0.9712	0.9842
0.02	0.9802	0.9802	0.9911	0.9493	0.9684
0.05	0.9520	0.9520	0.9766	0.9090	0.9242
0.1	0.9083	0.9083	0.9499	0.8501	0.8642
0.2	0.8319	0.8319	0.8931	0.7538	0.7693
0.5	0.6642	0.6643	0.7514	0.5681	0.5604
1	0.4973	0.4971	0.5882	0.4041	0.3979
2	0.3307	0.3302	0.4100	0.2547	0.2526
5	0.1646	0.1641	0.2145	0.2004	0.1197
10	0.0889	0.0893	0.1198	0.1118	0.0652
(b) Optimal damping ratio ζ					
0.01	0.0603	0.0547	0.0025	0.0655	0.0025
0.02	0.0841	0.0769	0.0065	0.0973	0.0073
0.05	0.1276	0.1199	0.0224	0.1477	0.0270
0.1	0.1686	0.1657	0.0505	0.2086	0.0593
0.2	0.2101	0.2244	0.0981	0.2919	0.1180
0.5	0.2402	0.3175	0.2012	0.4294	0.3047
1	0.2235	0.3894	0.2905	0.5359	0.4354
2	0.1749	0.4505	0.3779	0.6325	0.5498
5	0.1002	0.5057	0.4525	0.5163	0.6593
10	0.0581	0.5288	0.4804	0.5313	0.6841

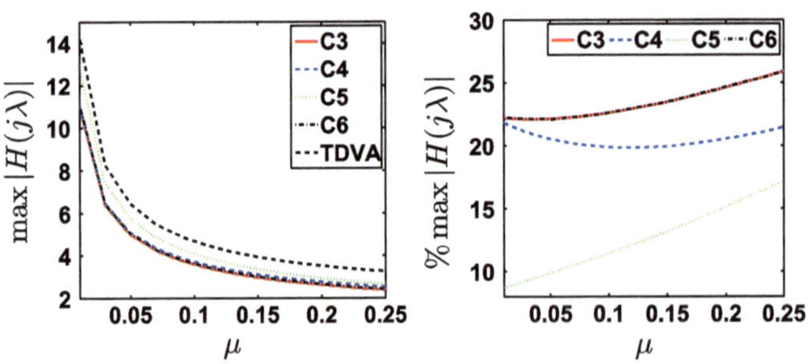

Fig. 4.8 Maximum magnitude comparison between the IDVAs and the TDVA (left figure) and percentage improvement of the IDVAs with respect to the TDVA (right figure)

Table 4.4 Optimal inertance-to-mass ratio δ and corner frequency ratio η in the H_∞ optimization

μ	$C3$	$C4$	$C5$	$C6$
(a) Optimal inertance-to-mass ratio δ				
0.01	0.0238	0.0234	2.2791	0.0228
0.02	0.0473	0.0453	1.8105	0.0448
0.05	0.1156	0.1069	1.6782	0.0989
0.1	0.2208	0.1930	1.5320	0.1538
0.2	0.4082	0.3212	1.1521	0.2126
0.5	0.8256	0.5719	0.6919	0.2426
1	1.2552	0.7785	0.3130	0.2009
2	1.7228	0.9703	0.1423	0.1364
5	2.2540	1.1307	3.9018	0.0627
10	2.4989	1.2089	3.6257	0.0339
(b) Optimal corner frequency ratio η				
μ	$C3$	$C4$	$C5$	$C6$
0.01	1.0051	0.9864	1.1242	1.0248
0.02	1.0098	0.9745	1.1982	1.0492
0.05	1.0248	0.9420	1.3341	1.1288
0.1	1.0485	0.9013	1.5181	1.2454
0.2	1.0940	0.8563	1.8754	1.4560
0.5	1.2219	0.7713	2.8856	2.2775
1	1.4061	0.7163	4.9686	3.5386
2	1.7178	0.6629	9.6074	6.0835
5	2.4169	0.6141	0.5009	14.5775
10	3.2632	0.5780	0.4739	27.6261

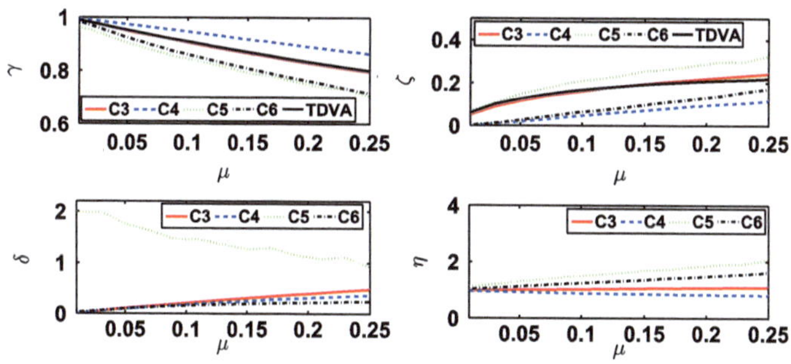

Fig. 4.9 Optimal parameters in the H_∞ optimization: natural frequency ratio γ (up left); damping ratio ζ (up right); inertance-to-mass ratio δ (bottom left); corner frequency ratio η (bottom right)

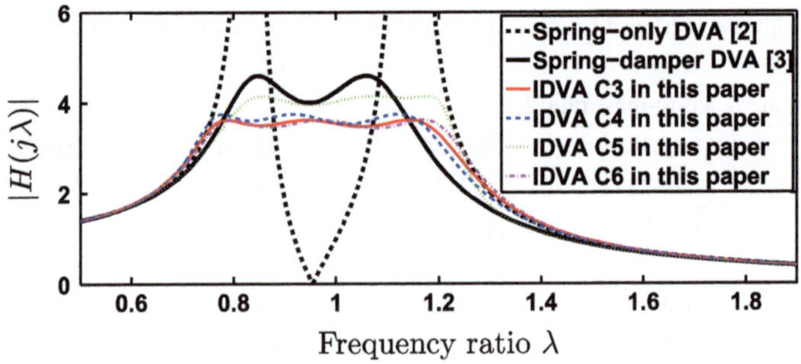

Fig. 4.10 Comparison between the IDVAs and TDVA when $\mu = 0.1$

4.5 H_2 Optimization for the IDVAs

4.5.1 H_2 Performance Measure and Its Analytical Solution

If the system is subjected to random excitation instead of sinusoidal excitation, the H_2 optimization would be more desirable than the H_∞ optimization (Asami et al. 1991, 2002; Cheung and Wong 2011a). The performance measure in the H_2 optimization is defined as (Asami et al. 1991, 2002; Cheung and Wong 2011a)

$$I = \frac{E\left[x^2\right]}{2\pi S_0 \omega_n}, \tag{4.11}$$

where S_0 is the uniform power spectrum density function. The mean square value of x of the object mass m can be calculated as

$$E\left[x^2\right] = S_0 \int_{-\infty}^{\infty} |H(j\lambda)|^2 d_\omega = S_0\omega_n \int_{-\infty}^{\infty} |H(j\lambda)|^2 d_\lambda, \tag{4.12}$$

where $H(j\lambda)$ is given in (4.7). Substituting (4.12) into (4.11), one obtains

$$I = \frac{1}{2\pi} \int_{-\infty}^{\infty} |H(j\lambda)|^2 d_\lambda, \tag{4.13}$$

which is exactly the definition of the H_2 norm of the transfer function $\hat{H}(s)$ by replacing $j\lambda$ in $H(j\lambda)$ with the Laplace variable s.

Therefore, the H_2 performance measure is rewritten as

$$I = \left\| \hat{H}(s) \right\|_2^2. \tag{4.14}$$

The analytical approach provided in Doyle et al. (1992, Chap. 2.6) will be employed to derive analytical solutions for IDVAs in the H_2 optimization, which is briefly presented as follows.

For a stable transfer function $\hat{H}(s)$, its H_2 norm can be calculated as (Doyle et al. 1992, Sect. 2.6)

$$\|\hat{H}(s)\|_2^2 = \|C(sI - A)^{-1}B\|_2^2 = CLC^T,$$

where A, B, and C are the minimal state-space realization $\hat{H}(s) = C(sI - A)^{-1}B$ and L is the unique solution of the Lyapunov equation

$$AL + LA^T + BB^T = 0. \tag{4.15}$$

We can write $\hat{H}(s)$

$$\hat{H}(s) = \frac{b_{n-1}s^{n-1} + \cdots + b_1 s + b_0}{s^n + a_{n-1}s^{n-1} + \cdots + a_1 s + a_0}$$

in its controllable canonical form below

$$\dot{x} = Ax + Bu, \ y = Cx,$$

where

$$A = \begin{bmatrix} 0 & 1 & 0 & \ldots & 0 \\ 0 & 0 & 1 & \ldots & 0 \\ \vdots & \vdots & \vdots & \vdots & \vdots \\ 0 & 0 & 0 & \ldots & 1 \\ -a_0 & -a_1 & -a_2 & \ldots & -a_{n-1} \end{bmatrix}, \ B = \begin{bmatrix} 0 \\ 0 \\ \vdots \\ 0 \\ 1 \end{bmatrix}, \ C = \begin{bmatrix} b_0, & b_1, & b_2 & \ldots & b_{n-1} \end{bmatrix}.$$

4.5.2 *Comparison Between the TDVA and IDVAs*

For the TDVA, the H_2 performance measure can be obtained as

$$I_{TDVA} = \frac{\gamma(1+\mu)\zeta}{\mu} + \frac{1 - (\mu+2)\gamma^2 + (1+\mu)^2\gamma^4}{4\mu\gamma\zeta}, \tag{4.16}$$

and the optimal γ and ζ are

$$\gamma_{TDVA,opt} = \sqrt{\frac{\mu+2}{2(1+\mu)^2}}, \tag{4.17}$$

$$\zeta_{TDVA,opt} = \sqrt{\frac{(3\mu+4)\mu}{8(\mu+1)(\mu+2)}}. \tag{4.18}$$

Substituting $\gamma_{TDVA,opt}$ and $\zeta_{TDVA,opt}$ into (4.16), one obtains the optimal $I_{TDVA,opt}$ as

$$I_{TDVA,opt} = \sqrt{\frac{3\mu+4}{4(\mu+1)\mu}}. \tag{4.19}$$

4.5.2.1 Performance Limitation of $C1$ and $C2$

The H_2 performance measures for $C1$ and $C2$ can be obtained as

$$I_{C1} = \frac{\gamma(1+\mu)\zeta}{\mu} + \frac{1}{4\mu\gamma\zeta}\left(\delta^2 - 2((1+\mu)\gamma^2-1)\delta + 1 - (\mu+2)\gamma^2 + (1+\mu)^2\gamma^4\right) \tag{4.20}$$

$$= I_{TDVA} + \frac{1}{4\mu\gamma\zeta}\left(\delta^2 + a_{C1,1}\delta\right), \tag{4.21}$$

$$I_{C2} = \left(a_{C2,2}\delta^{-2} + a_{C2,1}\delta^{-1} + a_{C2,0}\right)\zeta + \frac{1-(\mu+2)\gamma^2+(1+\mu)^2\gamma^4}{4\mu\gamma\zeta} \tag{4.22}$$

$$= I_{TDVA} + \left(a_{C2,2}\delta^{-2} + a_{C2,1}\delta^{-1}\right)\zeta, \tag{4.23}$$

where

$$\begin{aligned}
a_{C1,1} &= -2((1+\mu)\gamma^2-1),\\
a_{C2,2} &= \frac{\gamma}{\mu}\left((1+\mu)^3\gamma^4 - 2(1+\mu)\gamma^2 + 1\right),\\
a_{C2,1} &= \frac{\gamma}{\mu}\left(2+\mu-2(1+\mu)^2\gamma^2\right),\\
a_{C2,0} &= \frac{\gamma(1+\mu)}{\mu}.
\end{aligned}$$

The following proposition can be obtained.

Proposition 4.1 *For the H_2 performance, $C1$ performs no better than the TDVA.*

Proof See Appendix.

Proposition 4.2 *For the H_2 performance, $C2$ performs slightly better than the TDVA, but only at most 0.32% improvement can be achieved when $\mu \le 1$.*

Proof See Appendix.

Now, we have demonstrated that for the H_2 performance, $C1$ performs no better than the TDVA and $C2$ provides negligible improvement over the TDVA. This means that adding an inerter alone to the TDVA provides limited improvement for the H_2 performance, and therefore, another four IDVAs $C3$, $C4$, $C5$, and $C6$ are proposed by adding an inerter together with a spring to the TDVA. It will be shown in the following sections that in this way, the H_2 performance can be significantly improved.

4.5.2.2 Performance Benefits of $C3$, $C4$, $C5$, and $C6$

In this subsection, it will be analytically demonstrated that for the H_2 performance, IDVAs $C3$, $C4$, $C5$, and $C6$ perform surely better than the TDVA, and an optimization problem will be formulated to find the optimal parameters.

By using the method shown in Sect. 4.5.1, the analytical representations of the H_2 performance measures for $C3$, $C4$, $C5$, and $C6$ are calculated and the detailed equations are shown in Appendix. Denote the optimal H_2 performances of $C3$, $C4$, $C5$, and $C6$ as $I_{C3,opt}$, $I_{C4,opt}$, $I_{C5,opt}$, $I_{C6,opt}$, respectively. The following proposition can be obtained.

Proposition 4.3 *For the H_2 performance, IDVAs $C3$ and $C5$ always perform better than the TDVA, that is, the following inequalities hold:*

$$I_{C3,opt} < I_{TDVA,opt}, \tag{4.24}$$

$$I_{C5,opt} < I_{TDVA,opt}, \tag{4.25}$$

and if $\mu \leq 1$, IDVAs $C4$ and $C6$ always perform better than the TDVA, that is, the following inequalities hold:

$$I_{C4,opt} < I_{TDVA,opt}, \tag{4.26}$$

$$I_{C6,opt} < I_{TDVA,opt}, \tag{4.27}$$

where $I_{TDVA,opt}$ is the optimal H_2 performance for the TDVA given by (4.19).

Proof See Appendix.

Remark 6 *The condition $\mu \leq 1$ for $C4$ and $C6$ in Proposition 4.3 is only a sufficient condition, which means that for the case $\mu > 1$, it is also possible that the inequalities (4.26) and (4.27) hold. However, such a condition introduces no conservativeness for DVA applications, as the mass ratio μ is normally less than 1 in practice (typically less than 0.25) (Inman 2008; Cheung and Wong 2011b).*

Since the IDVAs $C3$, $C4$, $C5$, and $C6$ can always reduce to the TDVA by setting the spring stiffness k_1 (or η) and inertance b (or δ) to 0 or ∞, the conclusions $I_{Ci,opt} \leq$

$I_{TDVA,opt}$, $i = 3, 4, 5, 6$ *always hold. However, Proposition 4.3 demonstrates the existence of finite* η *and* δ *such that the IDVAs* $C3$, $C4$, $C5$, *and* $C6$ *are surely better than the TDVA.*

To determine the optimal values of δ, γ, η, and ζ, the following optimization problem should be solved.

$$\min_{\delta,\gamma,\eta,\zeta} I_{Ci}, i = 3, 4, 5, 6, \tag{4.28}$$

subject to $\delta > 0$, $\gamma > 0$, $\eta > 0$, and $\zeta > 0$.

Analytical solutions of $C3$: Problem (4.28) can be analytically solved for $C3$, where the optimal parameters for $C3$ are obtained as follows:

$$\gamma_{C3,opt} = \sqrt{\frac{\sqrt{17\mu^2 + 32\mu + 16} - \mu}{4(1+\mu)^2}}, \tag{4.29}$$

$$\eta_{C3,opt} = \sqrt{\frac{1 - 2(1+\mu)\gamma_{C3,opt}^2 + (1+\mu)\gamma_{C3,opt}^4}{(1 - (2+3\mu)\gamma_{C3,opt}^2 + (1+\mu)^2\gamma_{C3,opt}64)\gamma_{C3,opt}^2}}, \tag{4.30}$$

$$\delta_{C3,opt} = -\frac{2\hat{a}_{C3,2}}{\hat{a}_{C3,1}}, \tag{4.31}$$

$$\zeta_{C3,opt} = \sqrt{\frac{1 - (\mu+2)\gamma_{C3,opt}^2 + (1+\mu)^2\gamma_{C3,opt}^4}{4\mu\gamma_{C3,opt}(\hat{a}_{C3,2}\delta_{C3,opt}^{-2} + \hat{a}_{C3,1}\delta_{C3,opt}^{-1} + \hat{a}_{C3,0})}}, \tag{4.32}$$

where $\hat{a}_{C3,2}$, $\hat{a}_{C3,1}$, and $\hat{a}_{C3,0}$ are obtained by setting $\gamma = \gamma_{C3,opt}$ and $\eta = \eta_{C3,opt}$ for $a_{C3,2}$, $a_{C3,1}$, and $a_{C3,0}$, respectively. For the representations of $a_{C3,2}$, $a_{C3,1}$, and $a_{C3,0}$, see Appendix.

The analytical solutions δ, γ, and η are derived by successively setting the first derivatives of I_{C3} with respect to δ, η, and γ as 0, and then checking the sign of the second derivatives at stationary points. The optimal $\zeta_{C3,opt}$ is derived due to the fact that both parts on the right-hand side of (4.40) of I_{C3} are positive.

Solutions of $C4$, $C5$, **and** $C6$: The analytical solutions of $C4$, $C5$, and $C6$ cannot be obtained due to the high order equations (more than fourth order) involved in the derivation. However, the optimal solutions of η and ζ can be analytically represented with respect to δ and γ as follows:

$$\eta_{C4,opt} = \frac{\sqrt{-(g_{C4,1}\delta + f_{C4,1})(2f_{C4,2} + 2g_{C4,2}\delta + 2l_{C4,2}\delta^2)}}{2(f_{C4,2} + g_{C4,2}\delta + l_{C4,2}\delta^2)}, \tag{4.33}$$

$$\zeta_{C4,opt} = \sqrt{\frac{l_{C4,2}\eta^4\delta^2 + l_{C4,1}\delta + l_{C4,0}}{a_{C4,2}\delta^{-2} + a_{C4,1}\delta^{-1} + a_{C4,0}}}, \tag{4.34}$$

$$\delta_{C5,opt} = -\frac{2a_{C5,2}}{a_{C5,1}}, \tag{4.35}$$

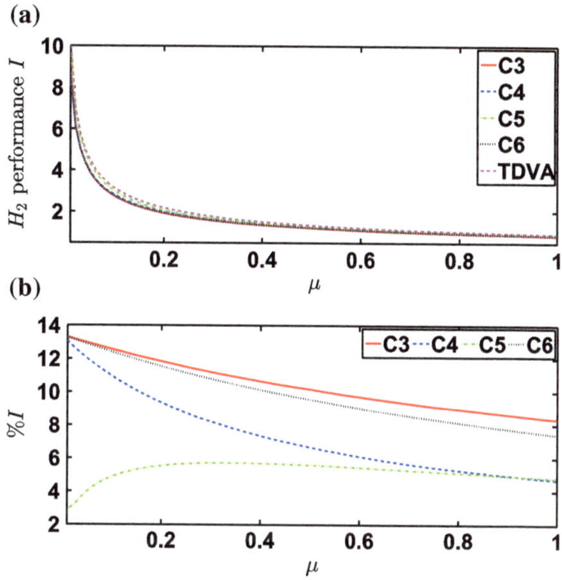

Fig. 4.11 Comparison between IDVAs and the TDVA. **a** the H_2 performance; **b** Percentage improvement of IDVAs with respect to the TDVA

$$\zeta_{C5,opt} = \sqrt{\frac{1-(\mu+2)\gamma^2+(1+\mu)^2\gamma^4}{4\mu\gamma(a_{C5,2}\delta_{C5,opt}^{-2}+a_{C5,1}\delta_{C5,opt}^{-1}+a_{C5,0})}}, \tag{4.36}$$

$$\zeta_{C6,opt} = \sqrt{\frac{l_{C6,2}\eta^4\delta^2+l_{C6,1}\delta+l_{C6,0}}{a_{C6,2}\delta^{-2}+a_{C6,1}\delta^{-1}+a_{C6,0}}}. \tag{4.37}$$

Correspondingly substituting the optimal representations above into I_{Ci}, $i = 4, 5, 6$, the problem (4.28) for Ci, $i = 4, 5, 6$ reduces to a nonlinear programming problem with two unknown variables δ and γ for $C4$ and $C5$, and with three unknown variables δ, γ and η for $C6$, which can be efficiently solved by using the Matlab solver *fmincon* and *GlobalSearch* in Global Optimization Toolbox.

Figures 4.11 and 4.12 depict the comparison between IDVAs $C3$, $C4$, $C5$, $C6$, and the TDVA when $0 \le \mu \le 1$. As shown in Fig. 4.11b, $C3$ performs the best, and more than 10% improvement with respect to the TDVA can be obtained by $C3$, $C4$ and $C6$. Similar to the H_∞ performance, the spring k_1 is better to be in series connection for the H_2 performance, given the fact that $C3$ and $C6$ are superior to $C4$ and $C5$.

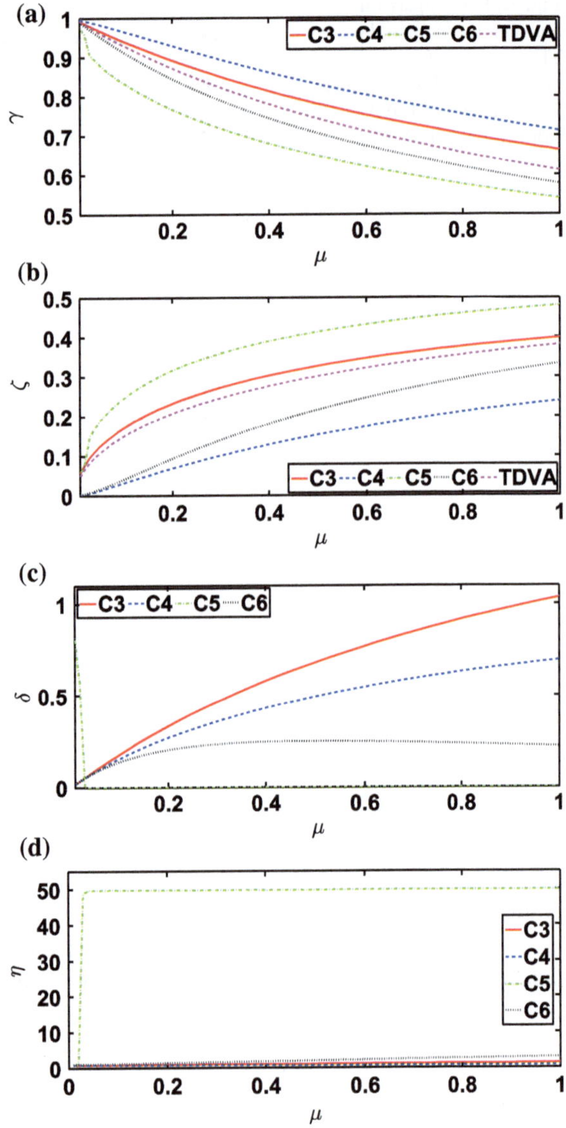

Fig. 4.12 Optimal parameters: **a** optimal γ; **b** optimal ζ; **c** optimal δ; **d** optimal η

4.6 Conclusions

In this chapter, the performance of inerter-based dynamic vibration absorbers (IDVAs) has been investigated, where the proposed IDVAs were a parallel arrangement of a spring and an inerter-based mechanical network. Both H_∞ and H_2 performances were considered. The H_∞ performance optimization problem was formulated in a minmax framework and solved by using a direct search optimization method;

while in the H_2 optimization, an analytical method was employed to calculate the H_2 performance measures. Comparisons between the proposed IDVAs and the traditional dynamic vibration absorber (TDVA) were conducted. The results showed that adding one inerter alone to the TDVA, no matter it is in parallel connection ($C1$) or in series connection ($C2$), provided no improvement for the H_∞ performance, and negligible improvement (less than 0.32% improvement over the TDVA when the mass ratio less than 1) for the H_2 performance. This demonstrated the necessity of introducing another degree of freedom together with the inerter to the TDVA, and then the IDVAs $C3$, $C4$, $C5$, and $C6$ were proposed by adding an inerter together with a spring to the TDVA. Significant improvement was obtained by IDVAs $C3$, $C4$, $C5$, and $C6$. For the H_∞ performance, numerical simulations showed that over 20% improvement was achieved compared with the TDVA and the effective frequency band can be enlarged by using inerter; while for the H_2 performance, it was analytically demonstrated that IDVAs $C3$, $C4$, $C5$, and $C6$ were surely better than the TDVA by carefully choosing the parameters, and over 10% improvement was obtained in the numerical simulation.

Appendix

Detailed representations of R_{ni}, I_{ni}, R_{mi}, and I_{mi}, $i = 1, \ldots, 6$.

$$
\begin{aligned}
R_{n1} &= \lambda^2 - \gamma^2 + \delta\lambda^2, \\
I_{n1} &= -2\lambda\gamma\zeta, \\
R_{m1} &= (-\mu\delta - \delta - 1)\lambda^4 + (\gamma^2 + \mu\gamma^2 + 1 + \delta)\lambda^2 - \gamma^2, \\
I_{m1} &= 2\lambda\gamma\zeta(\lambda^2 - 1 + \mu\lambda^2), \\
R_{n2} &= \delta\lambda(\gamma^2 - \lambda^2), \\
I_{n2} &= -2\gamma\zeta(\gamma^2 - (1 + \delta)\lambda^2), \\
R_{m2} &= \delta\lambda(\lambda^4 - (\gamma^2 + \mu\gamma^2 + 1)\lambda^2 + \gamma^2), \\
I_{m2} &= -2\gamma\zeta((1 + \delta + \mu\delta)\lambda^4 - (\gamma^2 + \mu\gamma^2 + 1 + \delta)\lambda^2 + \gamma^2), \\
R_{n3} &= \delta\eta^2\gamma\lambda(\gamma^2 - \lambda^2), \\
I_{n3} &= -2\zeta(\gamma^4\eta^2 - (1 + \delta\eta^2 + \eta^2)\lambda^2\gamma^2 + \lambda^4), \\
R_{m3} &= \delta\eta^2\gamma\lambda(\lambda^4 - (1 + \gamma^2 + \mu\gamma^2)\lambda^2 + \gamma^2), \\
I_{m3} &= 2\zeta(\lambda^6 - (1 + \mu + \eta^2 + \delta\eta^2 + \mu\delta\eta^2)\lambda^4 + ((\mu + 1)\eta^2\gamma^2 + 1 + \eta^2 + \\
&\quad \delta\eta^2)\gamma^2\lambda^2 - \gamma^4\eta^2, \\
R_{n4} &= -\delta(\lambda^4 - (1 + \eta^2 + \delta\eta^2)\gamma^2\lambda^2 + \gamma^4\eta^2), \\
I_{n4} &= -2\gamma\lambda\zeta(\gamma^2 - \lambda^2 - \delta\lambda^2), \\
R_{m4} &= \delta(\lambda^6 - (1 + (1 + \mu + \eta^2 + \delta\eta^2 + \delta\mu\eta^2)\gamma^2)\lambda^4 + ((\mu + 1)\eta^2\gamma^2 + (1 + \eta^2 + \\
&\quad \delta\eta^2))\gamma^2\eta^2 - \gamma^4\eta^2), \\
I_{m4} &= -2\gamma\lambda\zeta((1 + \delta + \mu\delta)\lambda^4 - (1 + \delta + \gamma^2 + \mu\gamma^2)\lambda^2 + \gamma^2),
\end{aligned}
$$

$$\begin{aligned}
R_{n5} &= \delta(\gamma^2 - \lambda^2)(\lambda^2 - \eta^2\gamma^2),\\
I_{n5} &= -2\gamma\lambda\zeta((1+\delta\eta^2)\gamma^2 - (1+\delta)\lambda^2),\\
R_{m5} &= \delta(\lambda^2 - \eta^2\gamma^2)(\lambda^4 - (1+\gamma^2+\mu\gamma^2)\lambda^2 + \gamma^2),\\
I_{m5} &= -2\gamma\lambda\zeta((1+\delta+\mu\delta)\lambda^4 - ((1+\mu+\delta\eta^2+\mu\delta\eta^2)\gamma^2 + 1 + \delta)\lambda^2 +\\
&\quad (1+\delta\eta^2)\gamma^2),\\
R_{n6} &= -\delta(\lambda^4 - (1+\eta^2+\delta\eta^2)\gamma^2\lambda^2 + \gamma^4\eta^2),\\
I_{n6} &= 2\lambda\gamma\zeta(\lambda^2 - (1+\delta\eta^2)\gamma^2),\\
R_{m6} &= \delta(\lambda^6 - (1+(1+\mu+\eta^2+\delta\eta^2+\mu\delta\eta^2))\lambda^4 + ((\mu+1)\eta^2\gamma^2 +\\
&\quad (1+\eta^2+\delta\eta^2))\gamma^2\lambda^2 - \gamma^4\eta^2),\\
I_{m6} &= -2\gamma\lambda\zeta(\lambda^4 - (1+(1+\mu+\delta\eta^2+\mu\delta\eta^2)\gamma^2)\lambda^2 + (1+\delta\eta^2)\gamma^2).
\end{aligned}$$

Proof of Proposition 4.1

From (4.21), if $C1$ performs better than the TDVA, that is $I_{C1} < I_{TDVA}$, the second term of (4.21) must be less than 0, which means

$$\delta^2 + a_{C1,1}\delta < 0.$$

Since $\delta \geq 0$, if $\gamma^2 < \frac{1}{1+\mu}$, the optimal δ denoted as δ_{opt} is 0. If $\gamma^2 \geq \frac{1}{1+\mu}$, the optimal $\delta_{opt} = (1+\mu)\gamma^2 - 1$, and it can be checked that the optimal γ is $\frac{1}{1+\mu}$ by substituting δ_{opt} into (4.21), which means that the optimal δ is also 0.

Proof of Proposition 4.2

First, we prove that $C2$ performs better than the TDVA, that is $I_{C2,opt} < I_{TDVA,opt}$, where $I_{C2,opt}$ denotes the optimal I_{C2}. From (4.23), if $C2$ performs better than the TDVA, the following inequality must hold:

$$a_{C2,2}\delta^{-2} + a_{C2,1}\delta^{-1} < 0,$$

which requires that

$$a_{C2,1} < 0 \text{ or } \gamma^2 > \frac{2+\mu}{2(1+\mu)^2},$$

as $a_{C2,2} \geq 0$ for any $\gamma \geq 0$. If $\gamma^2 > \frac{2+\mu}{2(1+\mu)^2}$, the optimal δ^{-1} is

$$\delta_{opt}^{-1} = -\frac{a_{C2,1}}{2a_{C2,2}},$$

and I_{C2} can be represented as

$$I_{C2} = \sqrt{\frac{(1-(2+\mu)\gamma^2 + (1+\mu)^2\gamma^4)(4(1+\mu)^2\gamma^2 - \mu)}{4\mu(1-2(1+\mu)\gamma^2 + (1+\mu)^3\gamma^4)}}. \tag{4.38}$$

Using $I_{TDVA,opt}$ given in (4.16), one obtains

$$I_{C2}^2 - I_{TDVA,opt}^2 = \frac{((\mu+1)\gamma^2-1)(2(\mu+1)^2\gamma^2-2-\mu)^2}{4\mu(1-2(\mu+1)\gamma^2+(\mu+1)^3\gamma^4)(\mu+1)},$$

Clearly, if $\gamma^2 < \frac{1}{1+\mu}$, then $I_{C2} < I_{TDVA,opt}$. Since $\frac{1}{1+\mu} > \frac{2+\mu}{2(1+\mu)^2}$, one can always find a γ such that $I_{C2} < I_{TDVA,opt}$. Since $I_{C2,opt} \le I_{C2}$, one obtains $I_{C2,opt} < I_{TDVA,opt}$.

Second, we graphically prove that only at most 0.32% improvement can be obtained by $C2$ when $\mu \le 1$. The optimal γ can be obtained by solving $\frac{\partial I_{C2}^2}{\partial \gamma^2} = 0$, which is equivalent to

$$(2\alpha^2\gamma^2-1-\alpha)(2\alpha^5\gamma^6+(\alpha^4-7\alpha^3)\gamma^4+(8\alpha^2-2\alpha^3)\gamma^2-3\alpha+1)=0, \tag{4.39}$$

where $\alpha = \mu + 1$. It is easy to check that (4.39) has two real positive solutions denoted as γ_1 and γ_2, $\gamma_1 < \gamma_2$, where

$$\gamma_1 = \sqrt{\frac{1+\alpha}{2\alpha^2}},$$

and $\gamma_1 < \gamma_2 < \sqrt{2}\gamma_1$. Also, γ_2^2 is the unique real solution of equation

$$2\alpha^5\gamma^6+(\alpha^4-7\alpha^3)\gamma^4+(8\alpha^2-2\alpha^3)\gamma^2-3\alpha+1=0,$$

and the optimal γ is γ_2.

For $0 \le \mu \le 1$, a graphical comparison with the TDVA is shown in Fig. 4.13, where it is clearly shown that at most 0.32% improvement is obtained for $C2$.

Analytical representations of the H_2 performance measures for $C3$, $C4$, $C5$, and $C6$

Denote I_{C3}, I_{C4}, I_{C5}, and I_{C6} as the H_2 performance measures for $C3$, $C4$, $C5$, and $C6$, respectively. The detailed representations are obtained as follows:

$$\begin{aligned} I_{C3} &= \left(a_{C3,2}\delta^{-2}+a_{C3,1}\delta^{-1}+a_{C3,0}\right)\zeta + \frac{1-(\mu+2)\gamma^2+(1+\mu)^2\gamma^4}{4\gamma\mu\zeta} \\ &= I_{TDVA} + \left(a_{C3,2}\delta^{-2}+a_{C3,1}\delta^{-1}\right)\zeta, \end{aligned} \tag{4.40}$$

$$\begin{aligned} I_{C4} &= \left(a_{C4,2}\delta^{-2}+a_{C4,1}\delta^{-1}+a_{C4,0}\right)\zeta + \left(l_{C4,2}\eta^4\delta^2+l_{C4,1}\delta+l_{C4,0}\right)\frac{1}{\zeta} \\ &= I_{TDVA} + \left(a_{C4,2}\delta^{-2}+a_{C4,1}\delta^{-1}\right)\zeta + \left(l_{C4,2}\eta^4\delta^2+l_{C4,1}\delta+f_{C4,2}\eta^4+f_{C4,1}\eta^2\right)\frac{1}{\zeta}, \\ I_{C5} &= \left(a_{C5,2}\delta^{-2}+a_{C5,1}\delta^{-1}+a_{C5,0}\right)\zeta + \frac{1}{4\gamma\mu\zeta}\left(1-(\mu+2)\gamma^2+(1+\mu)^2\gamma^4\right) \\ &= I_{TDVA} + \left(a_{C5,2}\delta^{-2}+a_{C5,1}\delta^{-1}\right)\zeta, \end{aligned} \tag{4.41}$$

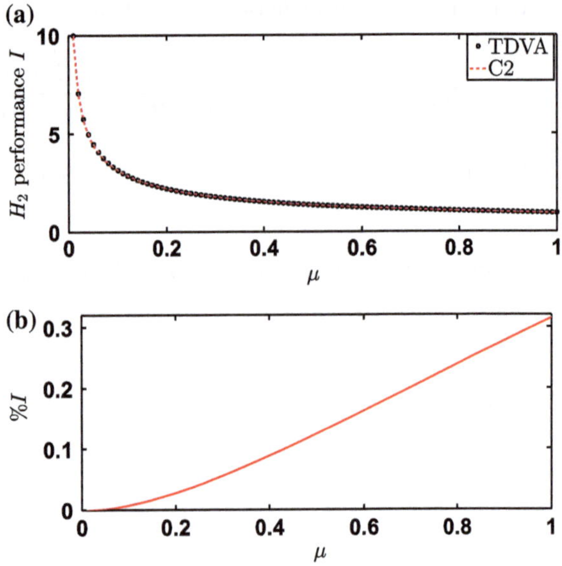

Fig. 4.13 Comparison between $C2$ and TDVA when $0 \le \mu \le 1$. **a** the H_2 performance; **b** Percentage improvement of $C2$ with respect to TDVA

$$
\begin{aligned}
I_{C6} &= \left(a_{C6,2}\delta^{-2}\eta^{-4} + a_{C6,1}\delta^{-1}\eta^{-2} + a_{C6,0}\right)\zeta + \left(l_{C6,2}\delta^{2} + l_{C6,1}\delta + l_{C6,0}\right)\frac{1}{\zeta} \\
&= I_{TDVA} + \left(a_{C6,2}\delta^{-2}\eta^{-4} + a_{C6,1}\delta^{-1}\eta^{-2}\right)\zeta + \\
&\quad \left(l_{C6,2}\delta^{2} + l_{C6,1}\delta + f_{C6,2}\eta^{-4} + f_{C6,1}\eta^{-2}\right)\frac{1}{\zeta},
\end{aligned}
$$

where

$$
\begin{aligned}
&a_{C3,2} = d_{C3,2}\eta^{-4} + d_{C3,1}\eta^{-2} + d_{C3,0}, \\
&a_{C3,1} = g_{C3,1}\eta^{-2} + g_{C3,0},\ a_{C3,0} = \frac{\gamma(1+\mu)}{\mu}, \\
&d_{C3,2} = \frac{1}{\gamma^3\mu}\left(1 - 2\gamma^2 + (1+\mu)\gamma^4\right),\ d_{C3,1} = -\frac{2}{\gamma\mu}\left(1 - (2+\mu)\gamma^2 + (1+\mu)^2\gamma^4\right), \\
&d_{C3,0} = \frac{\gamma}{\mu}\left(1 - 2(1+\mu)\gamma^2 + (1+\mu)^3\gamma^4\right),\ g_{C3,1} = -\frac{2}{\mu\gamma}\left(1 - (1+\mu)\gamma^2\right), \\
&g_{C3,0} = -\frac{\gamma}{\mu}\left(2(1+\mu)^2\gamma^2 - 2 - \mu\right), \\
&a_{C4,2} = \frac{\gamma}{\mu}\left(1 - (2+\mu)\gamma^2 + (1+\mu)^3\gamma^4\right), \\
&a_{C4,1} = \frac{\gamma}{\mu}\left(2 + \mu - 2(1+\mu)^2\gamma^2\right),\ a_{C4,0} = \frac{\gamma(1+\mu)}{\mu}, \\
&l_{C4,2} = \frac{\gamma^3(1+\mu)^2}{4\mu},\ l_{C4,1} = g_{C4,2}\eta^4 + g_{C4,1}\eta^2,\ l_{C4,0} = f_{C4,2}\eta^4 + f_{C4,1}\eta^2 + f_{C4,0}, \\
&g_{C4,2} = \frac{\gamma^3}{2\mu}\left(1 + \mu - (1+\mu)^3\gamma^2\right),\ g_{C4,1} = \frac{\gamma}{4\mu}\left(2(1+\mu)^2\gamma^2 - \mu - 2\right),
\end{aligned}
$$

$$f_{C4,2} = \frac{\gamma^3}{4\mu}\left((1+\mu)^4\gamma^4 + (\mu-2)(\mu+1)^2\gamma^2 + 1\right),$$
$$f_{C4,1} = -\frac{\gamma}{2\mu}\left((1+\mu)^3\gamma^4 - 2(1+\mu)\gamma^2 + 1\right),$$
$$f_{C4,0} = \frac{1}{4\mu\gamma}\left(1 - (\mu+2)\gamma^2 + (1+\mu)^2\gamma^4\right),$$
$$a_{C5,2} = \frac{g_{C5,2}\eta^4 + g_{C5,1}\eta^2 + g_{C5,0}}{\mu(1+f_{C5,1}\eta^2 + f_{C5,2}\eta^4)^2},$$
$$a_{C5,1} = \frac{l_{C5,3}\eta^6 + l_{C5,2}\eta^4 + l_{C5,1}\eta^2 + l_{C5,0}}{\mu(1+f_{C5,1}\eta^2 + f_{C5,2}\eta^4)^2},\ a_{C5,0} = \frac{\gamma(1+\mu)}{\mu},$$
$$g_{C5,2} = \gamma\left((1+\mu)\gamma^4 - 2\gamma^2 + 1\right),\ g_{C5,1} = -2\gamma\left((1+\mu)^2\gamma^4 - (\mu+2)\gamma^2 + 1\right),$$
$$g_{C5,0} = \gamma\left((1+\mu)^3\gamma^4 - 2(1+\mu)\gamma^2 + 1\right),\ f_{C5,1} = -(1+\gamma^2(1+\mu)),\ f_{C5,2} = \gamma^2,$$
$$l_{C5,3} = 2\gamma^3((1+\mu)^3 - 1),\ l_{C5,2} = -\gamma\left(4(1+\mu)^2\gamma^4 - 2\gamma^2 - \mu - 2\right),$$
$$l_{C5,1} = 2\gamma\left((1+\mu)^3\gamma^4 + (1+\mu)^2\gamma^2 - \mu - 2\right),\ l_{C5,0} = \gamma\left(\mu + 2 - 2(1+\mu)^2\gamma^2\right),$$
$$a_{C6,2} = \frac{1 - 2\gamma^2 + (1+\mu)\gamma^4}{\gamma^3\mu},\ a_{C6,1} = \frac{2((1+\mu)\gamma^2 - 1)}{\gamma\mu},\ a_{C6,0} = \frac{\gamma(1+\mu)}{\mu},$$
$$l_{C6,2} = \frac{1}{4\gamma\mu},\ l_{C6,1} = g_{C6,1}\eta^{-2} + g_{C6,0},$$
$$l_{C6,0} = f_{C6,2}\eta^{-4} + f_{C6,1}\eta^{-2} + f_{C6,0},$$
$$g_{C6,1} = \frac{\mu - 2 + 2\gamma^2}{4\gamma^3\mu},\ g_{C6,0} = \frac{1 - (1+\mu)\gamma^2}{2\gamma\mu},$$
$$f_{C6,2} = \frac{1 + (\mu-2)\gamma^2 + \gamma^4}{4\mu\gamma^5},\ f_{C6,1} = -\frac{1 - 2\gamma^2 + (1+\mu)\gamma^4}{2\mu\gamma^3},$$
$$f_{C6,0} = \frac{1 - (2+\mu)\gamma^2 + (1+\mu)^2\gamma^4}{4\gamma\mu}.$$

Proof of Proposition 4.3

For $C3$, substituting $\gamma_{TDVA,opt}$ and $\zeta_{TDVA,opt}$ into (4.40), one obtains

$$I'_{C3} = I_{TDVA,opt} + \left(a'_{C3,2}\delta^{-2} + a'_{C3,1}\delta^{-1}\right)\zeta_{TDVA,opt},$$

where $a'_{C3,2}$ and $a'_{C3,1}$ are obtained by setting $\gamma = \gamma_{TDVA,opt}$ for $a_{C3,2}$ and $a_{C3,1}$, respectively. It can be checked that $a'_{C3,2} > 0$ and

$$a'_{C3,1} = -\sqrt{\frac{2}{2+\mu}}\eta^{-2} < 0,$$

which means that there exist finite δ and η such that $I'_{C3} < I_{TDVA,opt}$. Since $I_{C3,opt} \leq I'_{C3}$, then one obtains $I_{C3,opt} < I_{TDVA,opt}$.

For $C4$, denote

$$I'_{C4} = 2\sqrt{\left(a'_{C4,2}\delta^{-2} + a'_{C4,1}\delta^{-1} + a'_{C4,0}\right)\left(l'_{C4,2}\eta^4\delta^2 + l'_{C4,1}\delta + l'_{C4,0}\right)},$$

where $a'_{C4,2}$, $a'_{C4,1}$, $a'_{C4,0}$, $l'_{C4,2}$, $l'_{C4,2}$, and $l'_{C4,0}$ are obtained by setting $\gamma = \gamma_{TDVA,opt}$.

Expanding I'_{C4}, one obtains

$$I'_{C4} = 2\sqrt{a'_{C4,0}f'_{C4,0} + f_{C4,\eta}}, \tag{4.42}$$

where

$$\begin{aligned} f_{C4,\eta} = &\left(l'_{C4,2}\delta^2 + g'_{C4,2}\delta + f'_{C4,2}\right)(a'_{C4,2}\delta^{-2} + \\ &a'_{C4,0})\eta^4 + f'_{C4,1}(a'_{C4,2}\delta^{-2} + a'_{C4,0})\eta^2 + f'_{C4,0}a'_{C4,2}\delta^{-2}. \end{aligned}$$

Note that

$$I_{TDVA,opt} = 2\sqrt{a'_{C4,0}f'_{C4,0}}.$$

Then, we will prove that there exist finite δ and η so that $f_{C4,\eta} < 0$. It can be checked that $l'_{C4,2}\delta^2 + g'_{C4,2}\delta + f'_{C4,2} > 0$, $a'_{C4,2}\delta^{-2} + a'_{C4,0} > 0$, and $f'_{C4,1}(a'_{C4,2}\delta^{-2} + a'_{C4,0}) < 0$. The discriminant of $f_{C4,\eta} = 0$ is

$$\begin{aligned} \Delta = (a'_{C4,2}\delta^2 + a'_{C4,0})\Big(&({f'_{C4,1}}^2 - 4f'_{C4,2}f'_{C4,0})a'_{C4,2}\delta^{-2} - 4g'_{C4,2}f'_{C4,0}a'_{C4,2}\delta^{-1} + \\ &{f'_{C4,1}}^2 a'_{C4,0} - 4l'_{C4,2}f'_{C4,0}a'_{C4,2}\Big). \end{aligned}$$

It can be checked that if $\mu < \frac{8\sqrt{2}-4}{7} \approx 1.045$, there exists a finite δ such that the second term of Δ is positive, which means that if $\mu < 1.045$, there exists a finite η such that $f_{C4,\eta} < 0$. For example, if choosing

$$\delta^{-1} = \frac{2g'_{C4,2}f'_{C4,0}}{{f'_{C4,1}}^2 - 4f'_{C4,2}f'_{C4,0}} = \frac{(3\mu+4)(1+\mu)}{4\mu(\mu+2)}, \tag{4.43}$$

and

$$\eta = \sqrt{\frac{-f'_{C4,1}}{l'_{C4,2}\delta^2 + g'_{C4,2}\delta + f'_{C4,2}}} = \sqrt{\frac{2(3\mu+4)^2(1+\mu)(4+\mu)}{(\mu+2)(43\mu^3 + 204\mu^2 + 272\mu + 64)}}, \tag{4.44}$$

one obtains

$$f_{C4,\eta} = \frac{1}{128} \frac{(7\mu^2 + 8\mu - 16)(\mu + 4)(3\mu + 4)^2}{\mu(43\mu^3 + 204\mu^2 + 272\mu + 64)(1 + \mu)(\mu + 2)} < 0.$$

From (4.42) and for the δ and η given by (4.43) and (4.44), one obtains that if $\mu < 1.045$,

$$I'_{C4} < I_{TDVA,opt}.$$

Since $I_{C4,opt} \leq I'_{C4}$, one obtains that if $\mu < 1.045$, $I_{C4,opt} < I_{TDVA,opt}$.

For $C5$, setting $\gamma = \gamma_{TDVA,opt}$ and $\zeta = \zeta_{TDVA,opt}$ in (4.41), one obtains

$$I'_{C5} = I_{TDVA,opt} + \left(a'_{C5,2}\delta^{-2} + a'_{C5,1}\delta^{-1}\right) \zeta_{TDVA,opt}. \tag{4.45}$$

Then, we will show that there exist finite δ and η such that $a'_{C5,2}\delta^{-2} + a'_{C5,1}\delta^{-1} < 0$. It can be checked that $a'_{C5,2} > 0$. Therefore, we only need to prove that there exists a finite η such that $a'_{C5,1} < 0$. Since

$$a'_{C5,1} = \frac{l'_{C5,3}\eta^6 + l'_{C5,2}\eta^4 + l'_{C5,1}\eta^2}{\mu(1 + f'_{C5,1}\eta^2 + f'_{C5,2}\eta^4)^2},$$

it is easy to check that $a'_{C5,1} < 0$ if $\eta^2 > (\mu + 1)\left(\mu + 1 + \sqrt{\mu^2 + 2\mu}\right)$ or $\eta^2 < (\mu + 1)\left(\mu + 1 - \sqrt{\mu^2 + 2\mu}\right)$. For example, if choosing

$$\eta = \sqrt{2(1 + \mu)^2}, \tag{4.46}$$

$$\delta^{-1} = \frac{2(2 + \mu)(\mu + 1)^2}{(1 + 8\mu + 4\mu^2)(4 + 9\mu + 4\mu^2)}, \tag{4.47}$$

one obtains

$$f_\delta = -\frac{\sqrt{2}(2 + \mu)^{5/2}(\mu + 1)^2}{(1 + 8\mu + 4\mu^2)(4 + 9\mu + 4\mu^2)(1 + 3\mu + 5\mu^2 + 2\mu^3)^2} < 0,$$

which means that for the η and δ given by (4.46) and (4.47), $I'_{C5} < I_{TDVA,opt}$. Since $I_{C5,opt} \leq I'_{C5}$, one obtains $I_{C5,opt} < I_{TDVA,opt}$.

For $C6$, setting $\gamma = \gamma_{TDVA,opt}$ and $\zeta = \zeta_{TDVA,opt}$, one obtains

$$I'_{C6} = I_{TDVA,opt} + f_{C6,\eta},$$

where $f_{C6,\eta} = d_2\eta^{-4} + d_1\eta^{-2} + d_0$, with

$$d_2 = a'_{C6,2}\zeta_{TDVA,opt}\delta^{-2} + f'_{C6,2}/\zeta_{TDVA,opt},$$
$$d_1 = a'_{C6,1}\zeta_{TDVA,opt}\delta^{-1} + (g'_{C6,1}\delta + f'_{C6,1})/\zeta_{TDVA,opt},$$
$$d_0 = (l'_{C6,2}\delta^2 + g'_{C6,0}\delta)/\zeta_{TDVA,opt}.$$

It can be checked that $d_2 > 0$ for any δ and if $\mu < \sqrt{2}$, $d_1 < 0$. Thus, it remains to prove that there exists a finite $\eta > 0$ such that $f_{C6,\eta} < 0$. This can be done by checking the discriminant of $f_{C6,\eta}$, which is

$$\begin{aligned}\Delta &= d_1^2 - 4d_2d_0 \\ &= 16(\mu-4)(\mu+1)^8\delta^4 - 16\mu(4\mu^3+11\mu^2+5\mu-4)(\mu+1)^4\delta^3 + \\ &\quad 8\mu^2(5\mu^2+21\mu+20)(\mu+1)^3\delta^2 + \mu^3(3\mu+4)^2.\end{aligned}$$

It is easy to see that there always exists a finite δ such that $\Delta > 0$. For example, if choosing

$$\delta = \frac{\mu(4\mu^3+11\mu^2+5\mu-4-\sqrt{6\mu^6+56\mu^5+253\mu^4+606\mu^3+799\mu^2+568\mu+176})}{2(\mu-4)(\mu+1)^4},$$

which is larger than 0 if $\mu < 4$, one obtains

$$\Delta = \mu^3(3\mu+4)^2 > 0.$$

Therefore, we can always find a η^{-2} between the two real positive solutions of $f_{C6,\eta} = 0$ such that $f_{C6,\eta} < 0$. A possible choice is $\eta^{-2} = -\frac{d_1}{2d_2}$. This means that if carefully choosing δ and η, the inequality $I'_{C6} < I_{IDVA,opt}$ holds. Since $I_{C6,opt} \leq I'_{C6}$, one obtains $I_{C6,opt} < I_{TDVA,opt}$.

References

Anh, N. D., & Nguyen, N. X. (2013). Design of TMD for damped linear strucutures using the dual criterion of equivalent linearization method. *International Journal of Mehcanical Sciences*, *77*, 164–170.

Asami, T., Wakasono, T., Kameoka, K., Hasegawa, M., & Sekiguchi, H. (1991). Optimum design of dynamic absorbers for a system subjected to random excitation. *JSME International Journal Series 3, Vibration, Control Engineering, Engineering for Industry*, *34*(2), 218–226.

Asami, T., Nishihara, O., & Baz, A. M. (2002). Analytical solutions to H_∞ and H_2 optimization of dynamic vibration absorbers attached to damped linear systems. *Journal of Vibration and Acoustics*, *124*(2), 284–295.

Bekdas, G., & Nigdeli, S. M. (2013). Mass ratio factor for optimum tuned mass damper strategies. *International Journal of Mechanical Sciences*, *71*, 68–84.

Brzeski, P., Pavlovskaia, E., Kapitaniak, T., & Perlikowski, P. (2014). The application of inerter in tuned mass absorber. *International Journal of Non-Linear Mechanics*, *70*, 20–29.

Chen, M. Z. Q., Hu, Y., Huang, L., & Chen, G. (2014). Influence of inerter on natural frequencies of vibration systems. *Journal of Sound and Vibration*, *333*(7), 1874–1887.

Cheung, Y. L., & Wong, W. O. (2009). H_∞ and H_2 optimizations of a dyanmic vibration absorber for suppressing vibrations in plates. *Journal of Sound and Vibration*, *320*, 29–42.

Cheung, Y. L., & Wong, W. O. (2011a). H_2 optimization of a non-traditional dynamic vibration absorber for vibration control of structures under random force excitation. *Journal of Sound and Vibration*, *330*(6), 1039–1044.

Cheung, Y. L., & Wong, W. O. (2011b). H-infinity optimization of a variant design of the dynamic vibration absorber-Revisited and new results. *Journal of Sound and Vibration*, *330*(16), 3901–3912.

Crandall, S. H., & Mark, W. D. (1963). *Random vibration in mechanical systems*. New York: Academic Press.

Den Hartog, J. P. (1985). *Mechanical vibrations*. New York, USA: Dover Publications, INC.

Doyle, J. C., Francis, B. A., & Tannenbaum, A. R. (1992). *Feedback control theory*. Oxford: Maxwell Macmillan Int.

Frahm, H. (1909). *Device for damping vibrations of bodies*. US Patent, No. 989958. 30.

Gao, H., Zhan, W., Karimi, H. R., Yang, X., & Yin, S. (2013). Allocation of actuators and sensors for coupled-adjacent-building vibration attenuation. *IEEE Transactions on Industrial Electronics*, *60*(12), 5792–5801.

Ghosh, A., & Basu, B. (2007). A closed-form optimal tuning criteiron for TMD in damped structures. *Structural Control and Health Monitoring*, *14*, 681–692.

Hixson, E. L. (1988). *Mechanical impedance*. In C. M. Harris (Ed.), *Shock and vibration handbook* (3rd ed.). New York: McGraw-Hill (ch. 10).

Hu, Y., Chen, M. Z. Q., Shu, Z., & Huang, L. (2015). Analysis and optimisation for inerter-based isolators via fixed-point theory and algebraic solution. *Journal of Sound and Vibration*, *346*, 17–36.

Inman, D. J. (2008). *Engineering vibration* (3rd ed.). NJ: Prentice-Hall.

Lazar, I. F., Neild, S. A., & Wagg, D. J. (2014). Using an inerter-based device for structural vibration suppression. *Earthquake Engineering and Structure Dynamics*, *43*(8), 1129–1147.

Marian, L., & Giaralis, A. (2014). Optimal design of a novel tuned mass-damper-inerter (TMDI) passive vibration control configuration for stochastically support-excited structural systems. *Probabilistic Engineering Mechanics*, *38*, 156–164.

Miguelez, M. H., Rubio, L., Loya, J. A., & Fernandez-Saez, J. (2010). Improvement of chatter stability in boring operations with passive vibration absorbers. *International Journal of Mechanical Sciences*, *52*(10), 1376–1384.

Nishihara, O., & Asami, T. (2002). Closed-form solutions to the exact optimizations of dynamic vibration absorbers (minimizations of the maximum amplitude magnification factors). *Journal of Vibration and Acoustics*, *124*(4), 576–582.

Ormondroyd, J., & Den Hartog, J. P. (1928). The theory of the dynamic vibration absorber. *ASME Journal of Applied Mechanics*, *50*, 9–22.

Pai, P. F., & Schulz, M. J. (2000). A refined nonlinear vibration absorber. *International Journal of Mechanical Sciences*, *42*(3), 537–560.

Shearer, J. L., Murphy, A. T., & Richardson, H. H. (1967). *Introduction to system dynamics*. Reading, MA: Addison-Wesley.

Si, Y., Karimi, H. R., & Gao, H. (2014). Modelling and optimization of a passive structural control design for a spar-type floating wind turbine. *Engineering Structures*, *69*, 168–182.

Smith, M. C. (2002). Synthesis of mechanical networks: The inerter. *IEEE Transactions on Automatic Control*, *47*(1), 1648–1662.

Zhan, W., Cui, Y., Feng, Z., Cheung, K. C., Lam, J., & Gao, H. (2013). Joint optimization approach to building vibration control via multiple active tuned mass dampers. *Mechatronics*, *23*(3), 355–368.

Chapter 5
Semi-active Inerter and Adaptive Tuned Vibration Absorber

Abstract This chapter presents a novel framework to realize the semi-active inerter, and proposes a novel semi-active-inerter-based adaptive tuned vibration absorber (SIATVA). The proposed semi-active inerter can be realized by replacing the fixed-inertia flywheel in the existing flywheel-based inerters with a controllable-inertia flywheel (CIF). Then, by using the proposed semi-active inerter, a SIATVA is constructed, and two control methods, that is the frequency-tracker-based (FT) control and the phase-detector-based (PD) control, are derived. The experimental results show that both the FT control and the PD control can effectively neutralize the vibration of the primary mass, although the excitation frequency may vary. The proposed SIATVA can also tolerate the parameter variation of the primary system. As a result, it can be applied to a variety of primary systems without resetting the parameters. The performance degradation by the inherent damping is also demonstrated.

Keywords Semi-active inerter · Adaptive tuned vibration absorber · Controllable-inertia flywheel · Physical embodiments · Experiments

5.1 Introduction

To date, three types of inerters have been proposed, including the rack-pinion inerter (Smith 2002; Chen et al. 2009), the ball-screw inerter (Wang and Su 2008; Chen et al. 2009) and the hydraulic (or fluid) inerter (Wang et al. 2011; Gartner and Smith 2011; Tuluie 2010). See Chap. 1 for details. Note that for the rack-pinion inerters (Smith 2002; Chen et al. 2009), the ball-screw inerter in Wang and Su (2008), Chen et al. (2009), and the hydraulic inerters in Wang et al. (2011), Gartner and Smith Tuluie (2011), Tuluie (2010), the inertance cannot be adjusted online. This means the inertance cannot be adapted according to the variation of the external disturbances and environmental conditions.

To make the inertance adjustable, in Tsai and Huang (2011), a variable-inertia device (VID) by using a magnetic planetary gearbox is proposed. The variable-inertance inerter is actually a semi-active device, similar to the semi-active dampers and semi-active springs, where parameters (inertance, damping coefficient, and

M. Z. Q. Chen and Y. Hu, *Inerter and Its Application in Vibration Control Systems*, https://doi.org/10.1007/978-981-10-7089-1_5

spring stiffness) can be adjusted online by consuming small amount of energy. From this point of view, we proposed the semi-active inerter concept in Chen et al. (2014), which is defined as the inerter whose inertance can be controlled online. Almost at the same time, other terminologies regarding variable-inertance inerter were independently proposed, such as the "adaptive inerter" in Li et al. (2014), Li et al. (2015), "inerter which enables changes of inertance" (Brzeski et al. 2015). In this chapter, we uniformly use the term "semi-active inerter" to represent the inerter whose inertance can be adjusted online. Moreover, note that the term "semi-active inerter" was first used in Zhang et al. (2010). However, the so-called semi-active inerter in Zhang et al. (2010) is not a variable-inertance inerter as defined in Chen et al. (2014) and this chapter, but a semi-active suspension with a passive inerter.

In terms of the physical embodiments of semi-active inerter, in Hu et al. (2017), we proposed a general framework to realize the semi-active inerter, that is the inertance can be controlled online by adjusting either the transmission ratio or the moment of inertia of the flywheel. From this point of view, the variable-inertia device in Tsai and Huang (2011) and the "inerter which enables changes of inertance" in Brzeski et al. (2015) are realized by adjusting the transmission ratio based on a magnetic planetary gearbox and a continuously variable transmission (CVT) with gear ratio control system, respectively. In contrast, in Hu et al. (2017), the method by adjusting the inertia of a flywheel is demonstrated based a ball-screw mechanism.

In this chapter, the general framework to realize the semi-active inerter will be introduced in detail. The proposed method is a general framework to realize semi-active inerters, which means that although the proposed framework is illustrated based on a ball-screw inerter, other types of semi-active inerters such as rack-pinion inerters and hydraulic inerters can similarly be constructed. Besides, a novel semi-active-inerter-based ATVA (SIATVA) is proposed. Two control frameworks for the SIATVA, that is, the frequency-tracker-based (FT) control and the phase-detector-based (PD) control, are proposed. The proposed control frameworks are also general frameworks in the sense that other frequency-tracking and phase-detection techniques can similarly be implemented.

5.2 Preliminary

The tuned vibration absorber (TVA) is an auxiliary spring-mass system connected to a host structure in order to suppress the vibration of the host structure. It is a classical vibration control device extensively used in many fields of civil and mechanical engineering (Den Hartog 1985). Note that although the TVA has the similar structure with the dynamic vibration absorber (DVA) or tuned mass damper (TMD) (see Chap. 4, they are normally applied in different situations. For the DVA or TMD, the spring-mass system is implemented to suppress the vibration of the host structure over a wide range of excitation frequencies, such as the inerter-based DVAs in Chap. 4; while the TVA is to suppress the vibration at a specific excitation frequency (Bonello 2011; Brennan 2006). Figure 5.1 shows the schematic of TVA, where the principle idea is

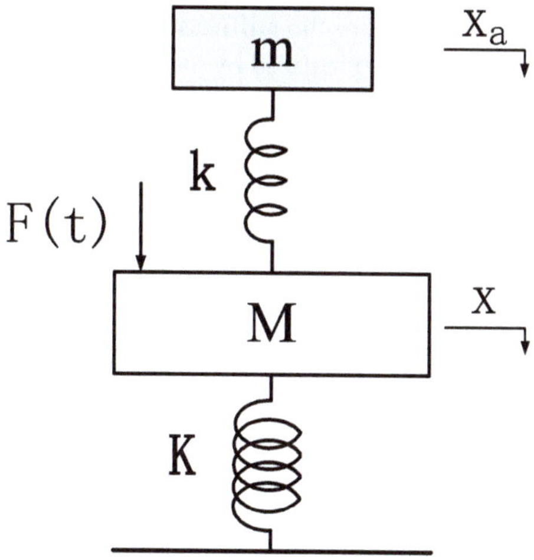

Fig. 5.1 The schematic of tuned vibration absorber (TVA), where K and M denote the host structure, k and m denote the spring-mass system

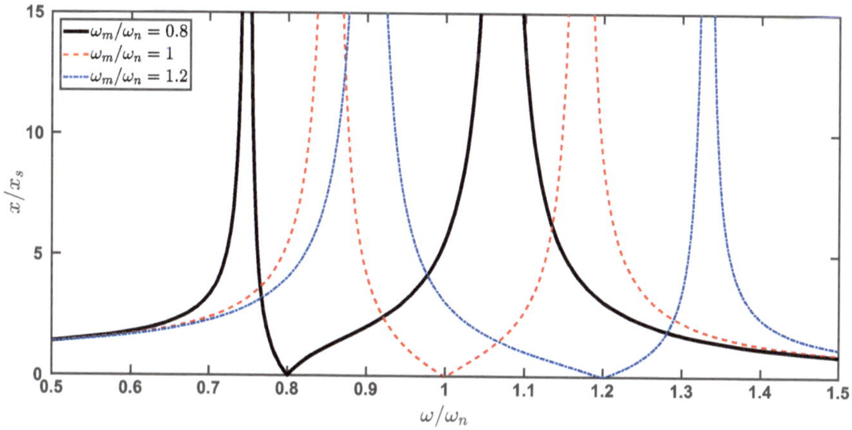

Fig. 5.2 The frequency response of the TVA with different ω_m, where $\omega_m = \sqrt{\frac{k}{m}}$, and $\omega_n = \sqrt{\frac{K}{M}}$

that if the damping of the auxiliary spring-mass system is neglectable, the vibration of the host structure at a specific frequency can be totally absorbed. As shown in Fig. 5.2 with different stiffness-mass pairs, if the stiffness and mass of the TVA are properly tuned, the vibration at a specific frequency can be totally suppressed.

Note that for the TVA, only the vibration at a specific frequency can be suppressed. This may be not applicable in practice as the environmental conditions may vary in time, and if the excitation frequency varies, the TVA would be not effective. To overcome this drawback, the adaptive TVA (ATVA) is developed, where the parameters (stiffness or mass) can be adapted online. A beam-like adjustable-stiffness TVA is proposed in Brennan (2006), where the stiffness is adjusted by moving the beams apart. Other types of ATVA can be realized by using shape memory alloy (Jayender et al. 2008), piezoceramic elements (Davis and Lesieutre 1999), etc. (see Bonello 2011; Brennan 2006) and references therein). In this chapter, another realization of ATVA by using the proposed semi-active inerter will be introduced and tested.

5.3 Semi-active Inerter

5.3.1 The Existing Inerters

Most of the existing inerters (Smith 2002; Chen et al. 2009; Wang and Su 2008; Wang et al. 2011) utilize a flywheel to realize the "inerter effect", and for all the existing flywheel-based inerters (Smith 2002; Chen et al. 2009; Wang and Su 2008; Wang et al. 2011), the inertance can be represented as the product of the square of a transmission ratio β and the moment of inertia of the flywheel J as follows

$$b = \beta^2 J. \tag{5.1}$$

For example, for the ballscrew inerter (Chen et al. 2009; Wang and Su 2008), β equals $(2\pi/p)$, where p is the pitch of the screw in units of $m/revolution$; for the rack-pinion inerter (Smith 2002; Chen et al. 2009), β is determined by the radiuses of the gears, pinions and the flywheel; for the hydraulic inerter in Wang et al. (2011), β is determined by the area of the piston.

Therefore, two means can be utilized to realize the semi-active inerter: the first one is to adaptively control the transmission ratio β online; the other is to adaptively control the moment of inertia of the flywheel J online. In this study, the latter is concerned and the method by replacing the fixed-inertia flywheel with a controllable-inertia flywheel (CIF) will be described in the following.

5.3.2 The Controllable-Inertia Flywheel (CIF)

In this study, a CIF is proposed based on the moving mechanical masses method (Schumacher 1991). Figure 5.3 shows the schematic of the proposed CIF. It involves at least two moving masses, which can radially move along the slots of the flywheel's main body. The slots are some channels enforcing the moving masses move radially and straightly. The moving masses are engaged with the collar through linkages, where the collar can move along the rotational axis. The position of the moving masses on the main body can be controlled by adjusting the position of the collar. An electric motor is mounted on the support structure to adjust the position of the collar,

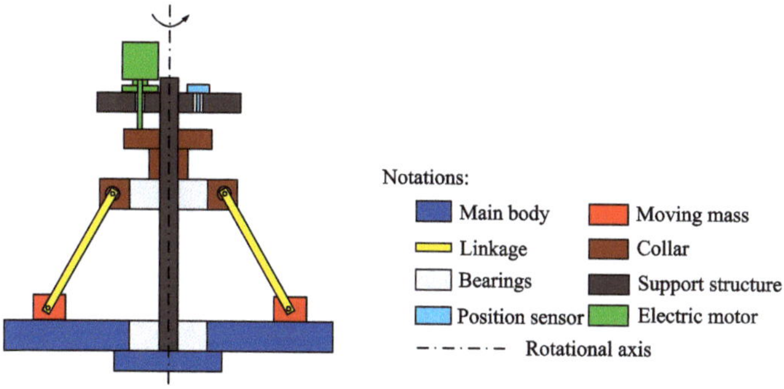

Fig. 5.3 The schematic of the proposed CIF

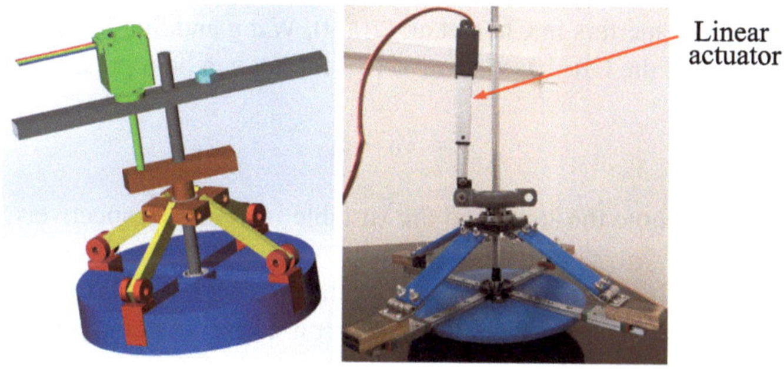

Fig. 5.4 The 3D representation (left) and a prototype (right) of the proposed CIF

and the support structure is fixed and cannot move in any direction. The position sensor is used to measure the distance between the support structure and the collar.

Different from the CIF in Schumacher (1991), in this chapter, two bearings are employed among the collar, the main body of the flywheel, and the support structure, as shown in Fig. 5.3. Thus, during rotation of the flywheel, the upper part of the collar and the support structure do not rotate with the flywheel, guaranteeing the smooth movement of the collar when pushed up and down by the electric motor. Figure 5.4 shows the 3D representation and a prototype of the proposed CIF with four moving masses, where in the prototype, the electrical motor and the position sensor have been embedded in the linear actuator.

Note that the number of moving masses can be adjusted if necessary. The moment of inertia of the CIF can be classified into two parts: the static part and the variable part. The static part includes the main body of the flywheel, the lower part of the collar, and the bearings, whose moment of inertia remains constant during rotation of the CIF. In contrast, the moving masses and the linkages constitute the variable part, whose moment of inertia can be controlled online.

Denote J, J_{static}, and $J_{variable}$ as the moment of inertia of the CIF, the static part, and the variable part, respectively. One has

$$J = J_{static} + J_{variable}, \tag{5.2}$$

where $J_{variable}$ is determined by the displacement of the linear actuator η. In this way, the moment of inertia of the CIF can be effectively controlled by varying the displacement of the linear actuator η.

5.3.3 *The CIF-Based Semi-active Inerter*

In this study, the proposed framework of realizing semi-active inerter is illustrated based on a ball-screw inerter, as shown in Fig. 5.5. Note that the CIF is located at the end of the screw to facilitate the operation of the CIF, which is different from the existing ball-screw inerters in Chen et al. (2009), Wang and Su (2008).

The inertance of the CIF-based semi-active inerter can be represented as

$$b = b_0 + b_v, \tag{5.3}$$

where b_0 and b_v denote the static and the variable inertance, respectively, and $b_0 = \beta^2 J_{static}$, $b_v = \beta^2 J_{variable}$.

The variable inertance b_v is determined by the displacement of the linear actuator η. Therefore, the inertance of the CIF-based semi-active inerter can be represented as

$$b = \Phi(\eta). \tag{5.4}$$

Note that $\Phi(\eta)$ is an increasing function with respect to η, which means that the minimal and maximal inertances of the CIF are $b_{\min} = \Phi(\eta_{\min})$ and $b_{\max} = \Phi(\eta_{\max})$, where $\eta_{\min}$ and $\eta_{\max}$ are the minimal and maximal η, respectively.

5.3.4 *Modeling of the Proposed Semi-active Inerter*

The proposed CIF-based semi-active inerter utilizes a linear actuator to adjust the displacement η, and the linear actuator is driven by a driven voltage V. Denoting the displacement η as $G(V)$, the dynamics of the CIF-based semi-active inerter can be

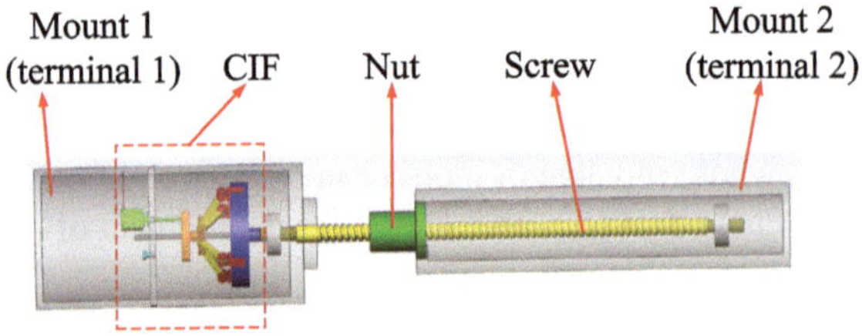

Fig. 5.5 An illustration of the proposed semi-active inerter based on a ballcrew inerter

summarized as

$$F = b(\ddot{x}_1 - \ddot{x}_2), \tag{5.5}$$

$$b = \Phi(\eta), \tag{5.6}$$

$$\eta = G(V). \tag{5.7}$$

The details on deriving $\Phi(\eta)$ will be introduced in Sect. 5.5.

5.4 Semi-active-Inerter-Based Adaptive Tuned Vibration Absorber

5.4.1 Problem Formulation

Figure 5.6 shows the comparison between the TATVA and the SIATVA, where M, K, C denote the mass, stiffness, and damping coefficient of the primary system, respectively. The object is to reject the harmonic force imposed on the primary mass, where the excitation frequency may vary with time.

Denote F as a sinusoidal force with a time-varying frequency. The transfer function from the disturbing force F to the displacement of the primary system for the SIATVA can be obtained as

$$T = \frac{x}{F} = \frac{bs^2 + k_1}{(Ms^2 + Cs + K)(bs^2 + k_1) + bk_1 s^2}. \tag{5.8}$$

As shown in (5.8), if the SIATVA is tuned to have $\omega = \sqrt{k_1/b}$, where ω is the excitation frequency of the force F, perfect cancellation can be achieved. Perfect cancellation means that the disturbing force F has no influence on the primary mass M. The perfect cancellation condition only involves the parameters of the SIATVA and does not rely on the parameters of the primary system.

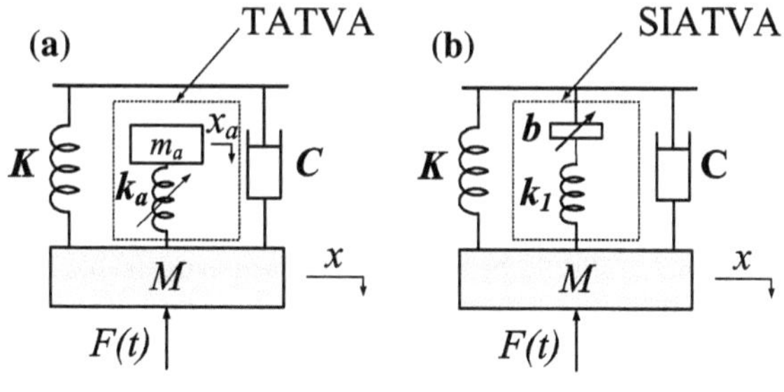

Fig. 5.6 The comparison of the traditional adaptive tuned vibration absorber (TATVA) and the semi-active inerter based adaptive tuned vibration absorber (SIATVA): **a** TATVA; **b** SIATVA

5.4.2 Frequency-Tracker-Based (FT) Control

According to the perfect cancellation condition, for a given stiffness k_1 and a given excitation frequency ω, the required inertance can be obtained as

$$b = \frac{k_1}{\omega^2}. \tag{5.9}$$

Therefore, an intuitive way to control the SIATVA is to track the excitation frequency online, and adjust the inertance according to (5.9). The frequency can be obtained by using a frequency tracker via the measurement of the acceleration of the primary mass. Therefore, only one sensor is required in the FT control.

Various techniques can be employed to track the frequency of harmonic signals online, such as zero-crossing detection (Friedman 1994), Kalman filters (Partovibakhsh and Liu 2014), etc. In this study, a zero-crossing-based frequency tracker is employed for its simplicity and efficiency in the experiments (for the details on zero-crossing detection, see (Friedman 1994) and references therein).

5.4.3 Phase-Detector-Based (PD) Control

Another framework to control the proposed SIATVA system is the phase-detector-based (PD) control. The proposed controller requires two measurements, i.e. the acceleration of the primary mass and the acceleration of the connection point between the spring k_1 and the inerter b.

Denote the displacement of the connection point as y. The transfer function from $\ddot{x}$ to $\ddot{y}$ can be obtained as $\ddot{y}/\ddot{x} = k_1/(k_1 + bs^2)$. Assume the excitation frequency is ω_0, and let $\ddot{x} = a_x \sin(\omega_0 t)$, where a_x is the amplitude, then $\ddot{y}$ can be represented as $\ddot{y} = a_y \sin(\omega_0 t - \phi)$, where $a_y = a_x k_1/(|k_1 - b\omega_0^2|)$, and

$$\phi = \begin{cases} 0 & \text{if } k_1/b > \omega_0^2, \\ \pi & \text{if } k_1/b < \omega_0^2, \\ \pi/2 & \text{if } k_1/b = \omega_0^2. \end{cases}$$

It is clear that the phase difference between $\ddot{x}$ and $\ddot{y}$ is $\pi/2$ when the perfect cancellation occurs. Therefore, the phase difference between ϕ and $\pi/2$ can be employed as the error signal, and a proper controller can be designed to minimize the phase difference. Note that if $\phi > \pi/2$, the inertance b should be decreased, and if $\phi < \pi/2$, the inertance b should be increased. It is also desirable for the designed controller to have large adjustments if the difference is significantly different from 0 and have small adjustments if it is close to 0. Since the inertance is adjusted by the displacement of the linear actuator η, a control law for η can be derived as

$$\eta_k - \eta_{k-1} = P_1\left(\frac{\pi}{2} - \phi_k\right) + P_3\left(\frac{\pi}{2} - \phi_k\right)^3 - D\dot{\phi}_k, \tag{5.10}$$

where k is the sample index, the cubic term is to make the controller have small adjustments if the phase difference is close to 0, and the derivative term is to improve the dynamic response of the system.

The phase angle ϕ can be estimated by the following equation (Brennan 2006)

$$\phi = \cos^{-1}\left(\frac{\frac{1}{T}\int_0^T \ddot{x}\ddot{y}dt}{\ddot{x}_{rms}\ddot{y}_{rms}}\right),$$

where T is the period of signals $\ddot{x}$ and $\ddot{y}$, and $\ddot{x}_{rms}$ and $\ddot{y}_{rms}$ are the root-mean-square (rms) values of the signal $\ddot{x}$ and $\ddot{y}$, respectively. For a sinusoidal signal, its rms value equals $1/\sqrt{2}$ of its magnitude.

5.5 Experimental Evaluation

In this section, the effectiveness of the proposed semi-active inerter and SIATVA will be experimentally verified.

5.5.1 Experimental Platform Description

As shown in Fig. 5.7, the disturbing force $F(t)$ imposed on the primary mass M is induced by the vibration shaker through the intermediate mass and intermediate spring; the sensors 1 and 2 are two accelerometers, where sensor 1 is used to measure the acceleration of the primary mass M, and sensor 2 is used to measure the acceleration of the point between the spring k_1 and the semi-active inerter b; the spring k_1 in Fig. 5.6 is realized by four parallel springs. The proposed FT control and PD control are realized by using SIMULINK. Displacements are obtained from accelerations by performing Fourier transformation and then conducting numerical integration.

The parameters of the experimental platform and the control methods are given in Table 5.1, where the parameters in the FT control and PD control are chosen by using the trail-and-error method. According to the perfect cancellation condition in Sect. 5.4, the effective frequency range for the proposed SIATVA is in $[\omega_{\min}, \omega_{\max}]$ where

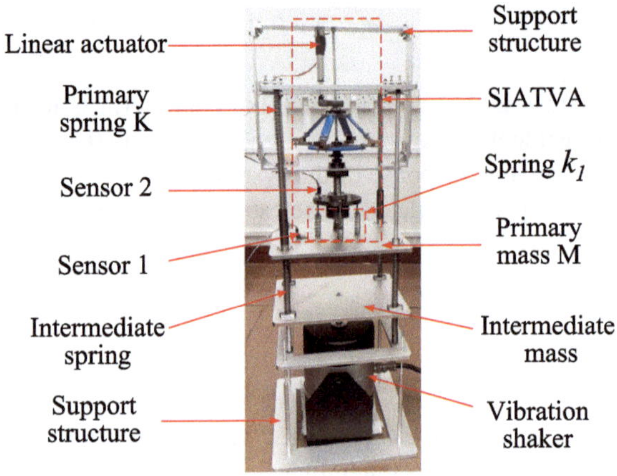

Fig. 5.7 The experimental platform for the proposed SIATVA

$$\omega_{\min} = \sqrt{\frac{k_1}{b_{\max}}} = 6.76 \text{ (rad/s)}, \tag{5.11}$$

$$\omega_{\max} = \sqrt{\frac{k_1}{b_{\min}}} = 12.10 \text{ (rad/s)}. \tag{5.12}$$

Note that compared with the existing ATVAs (Bonello 2011; Brennan 2006), the effective frequency range is quite small due to the limitation of the experimental setup. The approaches to enlarge the effective frequency range can be: enlarging the stiffness k_1, or enlarging the mass of the moving masses, or increasing the number of the moving masses, etc.

As shown in (5.6), the inertance b is determined by the displacement of the linear actuator η. The relation between b and η, i.e. the function $\Phi(\eta)$, can be obtained as follows: from Sect. 5.4, for a given excitation frequency ω, if the inertance satisfies the perfect cancellation condition, the response of the primary mass acceleration will be minimized. Since the spring stiffness k_1 can be measured, one can obtain a relation between b and η by measuring the specific η where minimal acceleration occurs for a given frequency ω, and then obtain b by $b = k_1/\omega^2$.

Figure 5.8 shows the relation between the inertance b and the linear actuator displacement η from both theoretical calculation and experimental measurement, where one sees that the theoretical calculation is consistent with the experimental measurement. The moment of inertia is a quadratic function of the rotary radius, and the rotary radius is a linear function of η. Therefore, a second-order polynomial is employed to fit the measured inertance and η in a least squares sense. The obtained curve fitting equation is

$$b = -0.0494\eta^2 + 6.5257\eta + 124.1974, \tag{5.13}$$

which will be employed as $\Phi(\eta)$ in the experiments in the FT control.

Table 5.1 The parameters of the experimental platform

	Description	Value
The primary system	Primary mass, M	2.5 kg
	Primary spring, K	896 N/m
	The intermediate mass	2.5 kg
	The intermediate spring	3576 N/m
Semi-active inerter	Number of moving masses, n	4
	Mass of per moving mass, m_{mm}	62.60 g
	Mass of the CIF main body, m_m	152.30 g
	Mass of the collar, m_c	60.20 g
	Mass of the bearing, m_b	24.90 g
	Mass of per linkage, m_L	3.9 g
	Pitch of the ball-screw, p	2 cm
	Min. rotary radius of moving masses	21.75 mm
	Max. rotary radius of moving masses	90.70 mm
	Min. linear actuator disp., $\eta_{\min}$	0 mm
	Max. linear actuator disp., $\eta_{\max}$	42 mm
	Static moment of inertia, J_{static}	0.5452 gm^2
	Static inertance from calculation, b_{static}	53.81 kg
	Min. inertance from calculation, $b_{\min}$	99.11 kg
	Max. inertance from calculation, $b_{\max}$	316.72 kg
	Speed of the linear actuator	15 mm/s
Spring k_1	Stiffness, k_1	14.5 kN/m
FT control	Sample time, T_s	10 ms
	Zero-crossing number per evaluation, N	25
PD control	Sample time, T_s	10 ms
	Linear proportional gain, P_1	2.5
	Cubic proportional gain, P_3	20
	Derivative gain, D	6

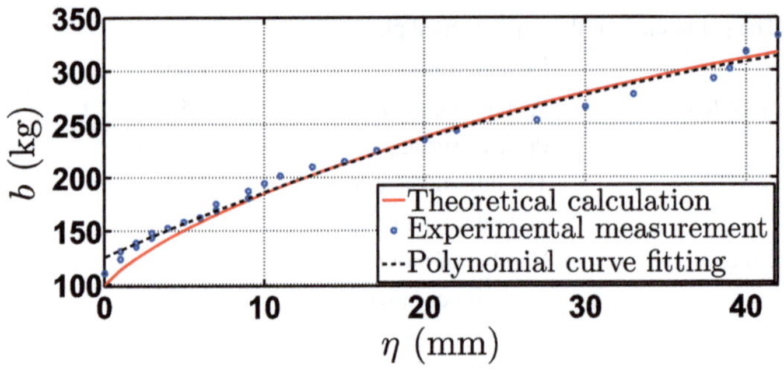

Fig. 5.8 Comparison of the relation between inertance b and linear actuator displacement η via theoretical calculation and experimental measurement

5.5.2 *Test Cases*

5.5.2.1 Test Case 1

This test is to demonstrate that the proposed SIATVA has the ability to adapt the inertance and thus to handle the excitation frequency variation by using the FT control and PD control. In the experiment, the excitation frequency ω varies as follows:

$$\omega = \begin{cases} 6.90 & t \in [0, 50], \\ 10.11 & t \in (50, 100], \\ 6.85 & t \in (100, 150]. \end{cases}$$

Figures 5.9 and 5.10 show the experimental results. Since some inherent nonlinearities exist in the experimental platform, such as the play in the ball screw, the friction among the components, etc., the perfect acceleration of the primary system cannot be achieved. However, compared with the case without the SIATVA, the proposed SIATVA can still significantly reduce the displacement of the primary mass. As shown in Fig. 5.10, both the FT control and PD control can automatically adjust the linear actuator displacement η to handle the variation of the excitation frequency.

Moreover, the experimental results show that the PD control tends to be more fluctuant than the FT control. The reason for this is that in this chapter, the frequency tracker tracks the frequency more stably than the phase detector detecting the phase, as shown in Fig. 5.10. Through experiments, it is also found that the FT control heavily relies on the performance of the frequency tracker, and if the disturbing force is too noisy (the high-order harmonics are too large), the FT control performs poorly. In this sense, the PD control is more robust to the variation of the disturbance than the FT control. However, all these issues can be alleviated by employing more efficient and more reliable frequency trackers and phase-detector-based controllers, and the main purpose of this test is to show that both of them are effective to control the SIATVA.

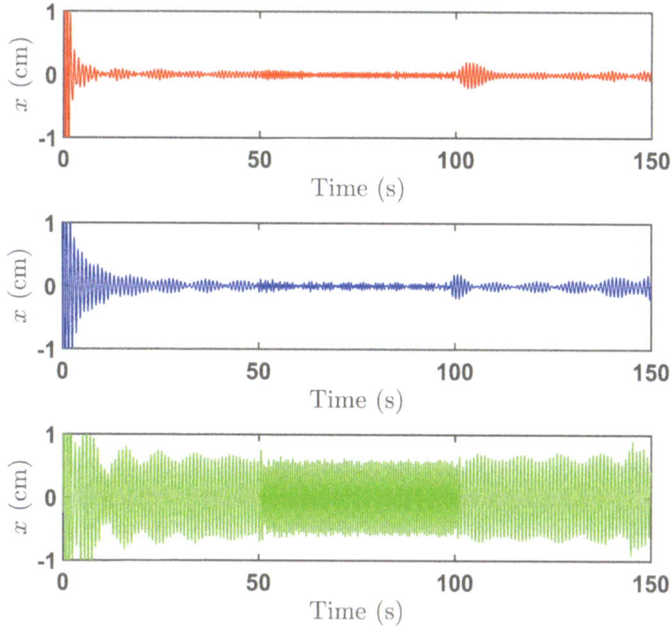

Fig. 5.9 The primary mass displacement in Test case 1: top: FT control; middle: PD control; bottom: without SIATVA

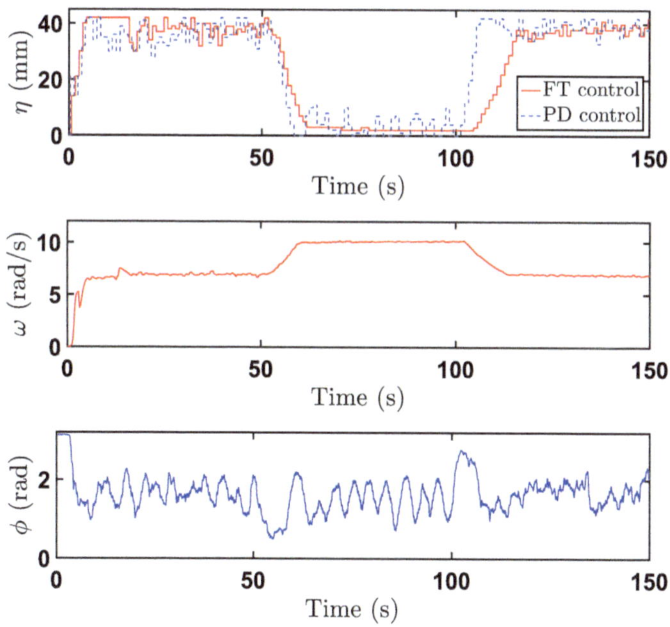

Fig. 5.10 Test case 1: the displacement of the linear actuator (top); the estimated frequency in FT control (middle); the phase difference in PD control (bottom)

5.5.2.2 Test Case 2

This test is to show the ability of the SIATVA to tolerate the variation of the primary system. The excitation frequency ω is 8.19 rad/s. The primary mass experiences a 43.6% sudden change, as represented as follows:

$$M = \begin{cases} 2.5 & t \in [0, 50], \\ 3.59 & t \in (50, 100], \\ 2.5 & t \in (100, 150]. \end{cases}$$

As shown in Fig. 5.11, the variation of the primary mass has a negligible influence on the performance of the SIATVA, which means that the proposed SIATVA can be used in a wide range of systems without resetting the parameters. Theoretically, if the inertance and the spring stiffness k_1 are well tuned, the SIATVA can be used in any primary systems. However, if the disturbing force is relatively much larger than the stiffness k_1, the free terminal of the semi-active inerter will undertake a very large vibration, which may lead to the stroke of the semi-active inerter to the limit.

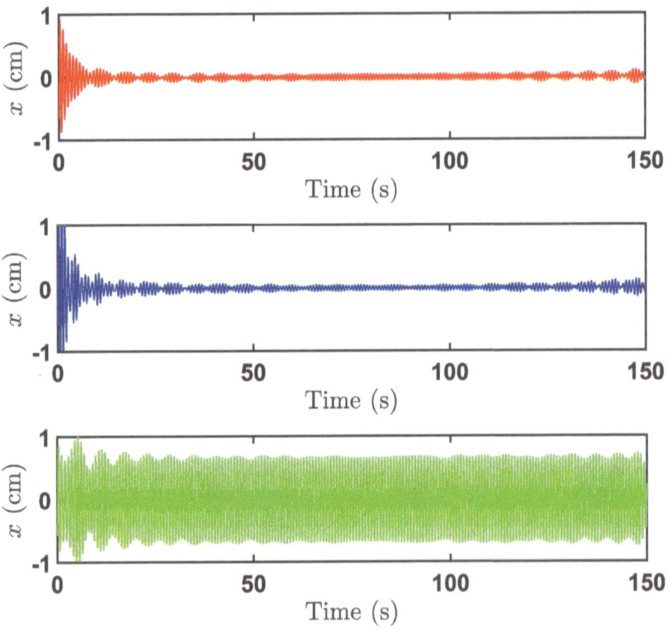

Fig. 5.11 The primary mass displacement in Test case 2: top: FT control; middle: PD control; bottom: without SIATVA

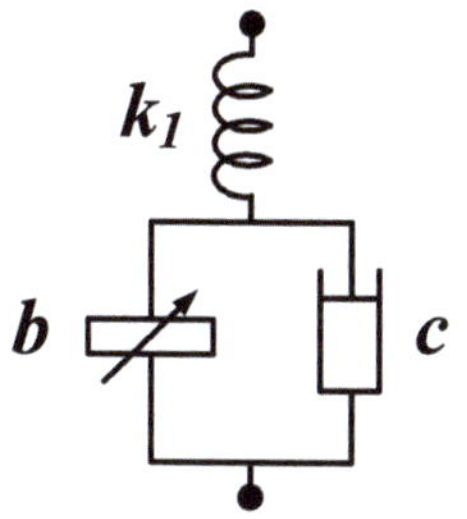

Fig. 5.12 The SIATVA with inherent damping

5.5.3 *The Influence of the Inherent Damping of the Semi-active Inerter*

The physical structure of the semi-active inerter, such as the friction, can introduce some inherent damping to the proposed semi-active inerter. In this study, the inherent damping is modeled as a viscous damper with a damping coefficient c. Then, the SIATVA can be modeled as shown in Fig. 5.12. The transfer function (5.8) can be rewritten as

$$\frac{x}{F} = \frac{bs^2 + cs + k_1}{(Ms^2 + Cs + K)(bs^2 + cs + k_1) + k_1(bs^2 + cs)},$$

where one sees that perfect cancellation cannot be achieved for a nonzero c. Figure 5.13 shows the frequency response of the primary mass's displacement with respect to different values of the inherent damping c, where the inertance b is set to be 200 kg. It is shown in Fig. 5.13 that the larger the inherent damping is, the worse the performance of the SIATVA will be, as increasing the inherent damping enlarges the displacement of the primary mass around the perfect cancellation frequency.

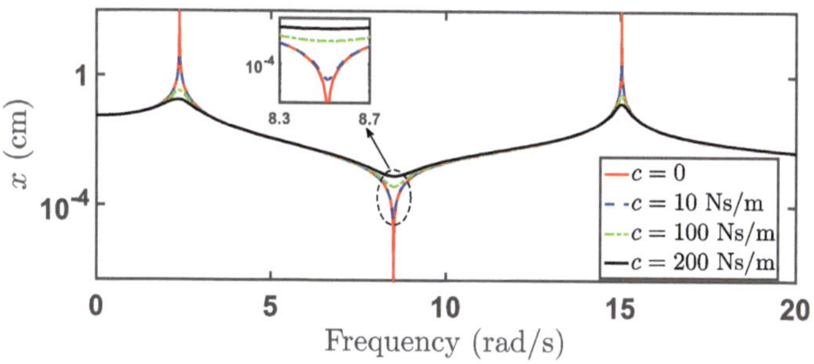

Fig. 5.13 The frequency response of the primary mass displacement with respect to different inerter inherent damping

5.6 Conclusions

In this chapter, a novel framework to realize the semi-active inerter and a novel semi-active-inerter-based adaptive tuneable vibration absorber (SIATVA) have been proposed. The proposed semi-active inerter can be obtained by replacing the fixed-inertia flywheel in the flywheel-based inerters with a controllable-inertia flywheel (CIF). Two frameworks to control the proposed SIATVA have been presented by using the frequency information (FT control) and the phase information (PD control), separately. Experimental results demonstrated that the proposed SIATVA with both the FT control and the PD control can effectively neutralize the vibration of the primary mass. The proposed SIATVA can also tolerate the parameter variation of the primary system, and can be applied to a variety of primary systems without resetting the parameters of the SIATVA. The fact that the inherent damping can degrade the performance of the SIATVA system was also demonstrated.

In terms of these two control methods, the FT control requires only one sensor while the PD control demands two. It was also found in the experiments that the FT control heavily relied on the performance of the frequency tracker, and the PD control was more fluctuant than the FT control. Therefore, it depends on the real applications to determine which method is more suitable in practice, and the main purpose of this chapter is just to show the effectiveness of these two methods in controlling the proposed SIATVA.

References

Bonello, P. (2011). Adaptive tuned vibration absorbers: Design principles. In F. Beltran-Carbajal (Ed.), *concepts and physical implementation, in Vibration Analysis and Control-New Trends and Developments* (pp. 1–26). Croatia: InTech.

Brennan, M. J. (2006). Some recent developments in adaptive tuned vibration absorbers/neutralisers. *Shock and Vibration, 13*, 531–543.

Brzeski, P., Kapitaniak, T., & Perlikowski, P. (2015). Novel type of tuned mass damper with inerter which enables changes of inertance. *Journal of Sound and Vibration, 349*(2015), 56–66.

Chen, M. Z. Q., Papageorgiou, C., Scheibe, F., Wang, F. C., & Smith, M. C. (2009). The missing mechanical circuit element. *IEEE Circuits and Systems Magazine, 9*(1), 10–26.

Chen, M. Z. Q., Hu, Y., Li, C., & Chen, G. (2014). *Semi-active suspension with semi-active inerter and semi-active damper, Proceedings of the 19th IFAC World Congress* (pp. 11225–11230). South Africa: Cape Town.

Davis, C., & Lesieutre, G. A. (1999). An actively tuned solid-state vibration absorber using capacitive shunting of piezoelectric stiffness. *Journal of Sound and Vibration, 232*(3), 601–617.

Den Hartog, J. P. (1985). *Mechanical vibrations*. New York, USA: Dover Publications Inc.

Friedman, V. (1994). A zero crossing algorithm for the estimation of the frequency of a single sinusoid in white noise. *IEEE Transactions on Signal Processing, 42*(6), 1565–1569.

Gartner, B. J., & Smith, M. C. (2011). Damper and inertial hydraulic device, U.S. Patent 13/577, 234.

Hu, Y., Chen, M. Z. Q, Xu, S., & Liu, Y. (2017). Semiactive inerter and its application in adaptive tuned vibration absorbers. *IEEE Transactions on Control Systems Technology, 25*(1), 294–300.

Jayender, J., Patel, R. V., Nikumb, S., & Ostojic, M. (2008). Modeling and control of shape memory alloy actuators. *IEEE Transactions on Control Systems Technology*, *16*(2), 279–287.

Li, P., Lam, J., & Cheung, K.-C. (2014). Investigation on semi-active control of vehicle suspension using adaptive inerter. *The 21st International Congress on Sound and Vibration*, 13–17 July 204, Beijing, China.

Li, P., Lam, J., & Cheung, K.-C. (2015). Control of vehicle suspension using an adaptive inerter. *Proceedings of the Institution of Mechanical Engineers, Part D: Journal of Automobile Engineering*, *229*(14), 1934–1943.

Partovibakhsh, M., & Liu, G. (2014). An adaptive Unscented Kalman Filtering approach for online estimation of model parameters and state-of-charge of lithium-ion batteries for autonomous mobile robots. *IEEE Transactions on Control Systems Technology*, *23*(1), 357–363.

Smith, M. C. (2002). Synthesis of mechanical networks: The inerter. *IEEE Transactions on Automatic Control*, *47*(1), 1648–1662.

Schumacher, L. L. (1991). Controllable inertia flywheel, U.S. Patent 4,995,282.

Tuluie, R. (2010). Fluid Inerter, U.S. Patent 13/575, 017.

Tsai, M.-C., & Huang, C.-C. (2011). Development of a variable-inertia device with a magnetic planetary Ggearbox. *IEEE/ASME Transactions on Mechatronics*, *16*(6), 1120–1128.

Wang, F.-C., & Su, W.-J. (2008). Impact of inerter nonlinearities on vehicle suspension control. *Vehicle System Dynamics*, *46*(7), 575–595.

Wang, F.-C., Hong, M. F., & Lin, T. C. (2011). Designing and testing a hydraulic inerter. *Proceedings of the Institution of Mechanical Engineers, Part C: Journal of Mechanical Engineering Science*, *225*(1), 66–72.

Zhang, X., Ahmadian, M., & Guo, K. (2010). A comparison of a semi-active inerter and a semi-active suspension. *SAE Technical Paper*, 2010-01-1903, https://doi.org/10.4271/2010-01-1903.

Chapter 6
Conclusions

As a new mechanical element, inerter has been applied in various mechanical systems. This book provides an extensive introduction of inerter, from the concept, the physical embodiments, the state of art research results, to the inerter-based vibration system analysis and design problems.

There are six chapters in this book. The first chapter presented a detailed description of inerter, its relation with network synthesis, the physical embodiments, and the current inerter-based vibration control systems. The inerter-based vibration control systems were divided into three categories, that is the passive, semi-active and active systems, based on the property of the control devices. The applications of inerter were reviewed, including the vehicle suspensions, train suspensions, tuned mass dampers (or dynamic vibration absorbers), building vibration control, aircraft landing gears, motorcycle steering compensators, rotors, bridges, etc. The state of art research on semi-active inerter was also reviewed.

In Chap. 2, the influence of inerter on natural frequencies of vibration systems was investigated. This is motivated by the fact that the performance benefits of inerter always demonstrated by using more complex structures comparing with the traditional spring-damper ones, where the fundamental property of inerter for vibration is not fully considered. Due to the importance of inerter, the influence of inerter was studied. It was theoretically proved that inerter can reduce the natural frequency of vibration systems. The position of the inerter on influencing the natural frequencies was also studied.

In Chap. 3, the isolating property of inerter was studied by proposing some inerter-based isolators. The investigation was conducted on two parts. The first part was to analyze the frequency response of two simple inerter-based networks, that is the parallel connection of an inerter and a damper, and the serious connection of an inerter and a damper. The influence of the simple inerter-based networks on the invariant points, the isolation properties was summarized. The second part was to present analytical parameter optimization methods for inerter-based isolators. The H_∞ and H_2 performance optimization methods were analytically derived.

M. Z. Q. Chen and Y. Hu, *Inerter and Its Application in Vibration Control Systems*, https://doi.org/10.1007/978-981-10-7089-1_6

In Chap. 4, the application of inerter in dynamic vibration absorption systems was studied, where inerter-based dynamic vibration absorbers (DVA) were proposed. The performances including the H_∞ and H_2 performances were evaluated numerically. It was demonstrated that only adding one inerter along to the traditional spring-damper DVA can only provide negligible performance improvements. However, if an inerter and a spring were added to the traditional DVA, the H_∞ and H_2 performances can be significantly improved. The comparison between inerter-based DVA and the traditional spring-damper DVA was conducted.

In Chap. 5, the semi-active inerter was introduced, and the general methods to realize semi-active inerter were presented. A physical embodiment of semi-active inerter by using a controllable-inertia flywheel was proposed and experimentally tested. The application of semi-active inerter in adaptive tuned vibration absorbers was considered. Control methods for the semi-active inerter were proposed, and the performance of the semi-active inerter-based adaptive tuned vibration absorbers was experimentally evaluated.

In this book, the recent advances in inerter and its related vibration control problems were introduced. Especially, the inerter-based isolation system, inerter-based dynamic vibration absorber, the semi-active inerter, and semi-active inerter-based adaptive tuned vibration absorber were analyzed in detail. Apart from the application of inerter in vibration control systems, inerter has potentially significant implications for theoretical and applied research in a broad range of scientific and engineering areas, such as systems and control fields. Over the longer term, the inerter also provides researchers with new methodologies and algorithms for application or evaluation in such diverse areas as passivity-preserving model reduction, open and interconnected systems, the passivity-based stability criterion, analysis and design of robust control, dissipative systems, linear-quadratic-Gaussian control of passive networks, high-precision synthesis, acoustic research, elastodynamic networks, microwave circuit design, filter deign, the design of electric ship control surface actuation systems, biometric image processing, and material science.

MIX
Papier aus verantwortungsvollen Quellen
Paper from responsible sources
FSC® C105338

If you have any concerns about our products,
you can contact us on
ProductSafety@springernature.com

In case Publisher is established outside the EU,
the EU authorized representative is:
Springer Nature Customer Service Center GmbH
Europaplatz 3, 69115 Heidelberg, Germany

Printed by Libri Plureos GmbH
in Hamburg, Germany